Shady Practices

CALIFORNIA STUDIES IN CRITICAL HUMAN GEOGRAPHY
Editorial Board

Shady Practices

*Agroforestry and Gender Politics
in The Gambia*

Richard A. Schroeder

UNIVERSITY OF CALIFORNIA PRESS
Berkeley · *Los Angeles* · *London*

Library of Congress Cataloging-in-Publication Data

Schroeder, Richard A.
    Shady practices : agroforestry and gender politics in the Gambia / Richard A.
Schroeder.
        p.   cm. — (California studies in critical human geography ; 5)
    Includes bibliographical references and index.
    ISBN 0-520-21687-3 (alk. paper)
        1. Mandingo (African people)—Agriculture.   2. Mandingo (African people)—So-
cial conditions.   3. Women, Mandingo—Economic conditions.   4. Division of labor—Gambia—Alkalikunda.   5. Patriarchy—Gambia—Alkalikunda.   6. Agroforestry—Gambia—Alkalikunda.   7. Sex role—Political aspects—Gambia—
Alkalikunda.   8. Forest ecology—Gambia—Alkalikunda.   9. Alkalikunda (Gambia)—Social life and customs.   I. Title.   II. Series.
DT509.45.M34S37   1999
338.1'096651—dc21   99-18198
                                                                                CIP

Manufactured in the United States of America

This book is a print-on-
demand volume. It is
manufactured using toner in
place of ink. Type and images
may be less sharp than the same
material seen in traditionally
printed University of California
Press editions.

The paper used in this publication meets the
minimum requirements of ANSI / NISO Z39.48-1992
(R 1997) (*Permanence of Paper*).

Portions of this text have been previously published. Chapter 3 first appeared as "'Gone
to their second husbands': Marital metaphors and conjugal contracts in The Gambia's
female garden sector," *Canadian Journal of African Studies* 30, 1 (1996): 69–87, and
has been reprinted with permission.

Excerpts of the following articles have also been reprinted with permission: "Shady
practice: Gender and the political ecology of resource stabilization in Gambian
garden / orchards," *Economic Geography* 69, 4 (1993): 349–365; "Contradictions
along the commodity road to environmental stabilization: Foresting Gambian gardens,"
Antipode 27, 4 (1995): 325–342; and "'Re-claiming' land in The Gambia: Gendered
property rights and environmental intervention," *Annals of the Association of American
Geographers* 87, 3 (1997): 487–508.

Tables 9 and 10 originally appeared in Susan B. Roberts, A. A. Paul, T. J. Cole, and
R. G. Whitehead "Seasonal changes in activity, birth weight and lactational performance in rural Gambian women," *Transactions of the Royal Society of Tropical Medicine and Hygiene* 76, 5 (1982): 668–678, and were reprinted in H. R. Barrett and A. W.
Browne "Time for development? The case of women's horticultural schemes in rural
Gambia," *Scottish Geographical Magazine* 105, 1 (1989): 4–11. Adapted versions of
the originals are reprinted here with permission.

Table 12 originally appeared in H. R. Barrett and A. W. Browne "Time for development? The case of women's horticultural schemes in rural Gambia," *Scottish Geographical Magazine* 105, 1 (1989): 4–11, and is reprinted with permission.

All photographs in this book are by the author.

This book is dedicated to the memory
of my mother,
Iola (Odie) Belle Root Schroeder Borchert
(1929–1978)

and my grandmother,
Esther Gertrude Margaret Gehring Root
(1900–1996)

Contents

Illustrations and Tables

Abbreviations

ANR	Agriculture and Natural Resources
CIDA	Canadian International Development Agency
DCD	Department of Community Development
EC	European Community
EEC	European Economic Community
ERP	Economic Recovery Program
FAO	United Nations Food and Agricultural Organization
GAD	Gender and Development
MCH	Maternal and Child Health Care
MID	MacCarthy Island Division
MMAP	Methodist Mission Agricultural Program
MRC	Medical Research Council
NBD	North Bank Division
NGO	Nongovernmental organization
SCF	Save the Children Federation
SIDA	Swedish International Development Agency

UNDP	United Nations Development Programme
UNICEF	United Nations International Children's Emergency Fund
USAID	United States Agency for International Development
VDC	Village Development Committee
WAD	Women and Development
WID	Women in Development

Preface

In 1986, I was hired by the Gambian field office of the U.S.-based non-governmental organization Save the Children Federation (SCF) to join the staff of a rural development program operating on The Gambia's North Bank. When I arrived in June of that year, the rainy season was about to begin and the farming season was imminent. Given that my duties were to oversee the agency's "food production sector," I quickly settled into the small town of Kerewan, where the agency's field coordinators were posted, and began work. My initial project assignments were both connected in some way to SCF drought-relief programs. The first involved distributing inputs such as seed, fertilizer, and tools to men's and women's groups for use on community farms, and the second entailed digging irrigation wells for several groups of women market gardeners. Grain grown by the community groups provided short-term relief to needy families whose crops had failed, whereas the cash-crop vegetable gardens helped rural families fundamentally restructure their livelihood systems in response to drought-induced changes in North Bank growing conditions.

In theory, both of these projects encouraged cooperation at the village level through labor exchanges and other forms of mutual support. In practice, however, as I soon discovered, they sometimes served to heighten social tensions among different community groups. Group identity in Mandinka society is often determined on the basis of membership in age cohorts (Mandinka: *kafolu*). For certain community tasks,

however, all but the most junior age sets gather together along gender lines, and three main groups are established—the men's group, the
women's group, and the youth group (a single group of unmarried
young men and women). This division of labor formed the basis for most
community development efforts organized by nongovernmental organization (NGOs) and state extension services in the 1980s, including one
of the projects I helped initiate, a conflict-ridden effort to provide wells
for a new women's market garden in the Upper Baddibu District community of Alkalikunda. In this project, four wells were to be dug by
a team of semi-skilled male laborers who were Alkalikunda residents.
The burden of the "community contribution" was to be shared by the
women's and men's groups according to a formula that was quite typical of NGO-sponsored construction projects in The Gambia at the time:
women were to provide sand and water, which were available locally
and could be transported using headpans; and men were to provide
gravel, which they had to collect with donkey carts from a gravel pit several kilometers away. Each of these components was mixed with cement
to form concrete caissons used to line the wells.

While the first of the four wells was completed without a hitch, construction efforts on the remaining wells hit a stumbling block when the
men's gravel contributions slowed to a trickle and finally stopped altogether. As the men explained, they saw no purpose in providing labor
for a project that would benefit the women at the men's expense. Moreover, since members of the well-digging team were being paid for their
efforts, the rest of the men felt it was unfair that they themselves should
work for free. The slowdown was understandably frustrating for the
women's group leaders, who were anxious to establish a garden nearer
to the village than their traditional garden site several kilometers away.
Construction of wells on the new site would give them a permanent supply of water and reduce the time spent in transit to and from their gardens. Thus, rather than allow the project to come to a halt, the women
organized an effort to gather the gravel themselves, painstakingly toting
individual headloads to the site until the wells were completed. This
gravel collection task caused tensions to emerge among the women,
however, and they carefully monitored each other to ensure that each
person upheld her share of the additional labor burden. It also height-
ened the women's resentment toward the members of the men's group
who had defaulted on their "community" commitments.

Several weeks into this stalemate, as I was meeting with the head of
the men's group on an unrelated matter in his groundnut (peanut) field

several hundred meters outside town, we looked up to see a delegation from the women's group marching across the field toward us. When they reached us, the leader of the group issued a perfunctory greeting and launched immediately into the business at hand. She demanded to know whether it was true, as the men's group leader had evidently claimed, that I had *ordered* the women to thresh the millet the men's group had harvested from their communal farm. All eyes on me now, I repeated the question so as to make sure I understood it and turned to the men's group leader for some sort of explanation. When he would not meet my gaze, I explained that I had no idea what the women were talking about and that I would never commit the women to such a task without discussing it with them directly first. The women then triumphantly turned to the men's group leader and said, "We *told* you! You can forget about us *ever* threshing that millet now. Thresh it yourself, for all we care, because we'll *never* do it for you!" With that, they turned on their heels and marched back to town. When the women had gone, the men's group leader demanded to know why I had failed to cover for him in his effort to get women to thresh grain for his group. When I challenged him in turn to explain why he had lied to the women about my having ordered the women's cooperation, he had no further reply and could only shake his head in disbelief.

This passing exchange in the Alkalikunda groundnut field revealed several important dimensions of the social dynamics surrounding rural development efforts on the North Bank which gave me pause for reflection at the time and subsequently gave rise to the study I describe in this book. First, the confrontation established clearly that the way we had chosen to implement SCF's integrated community development model failed to take into account some of the complexity of community-level politics in The Gambia. Community groups were often thoroughly cooperative when working on SCF projects, but, as the Alkalikunda exchange illustrates, this cooperation could neither be taken for granted nor imposed on people. In general, men's and women's groups seemed to see less joint benefit in their respective projects than we had assumed. In some cases, it seemed that our project assistance served to aggravate social divisions at the community level rather than promote community cohesion.[1]

Second, it was quite clear from the discussion in the groundnut field that the different groups of Alkalikunda residents were attempting to manipulate me and, by implication, the rest of the SCF staff, for political and economic advantage. While I was a conduit for development aid

and had considerable leverage over local groups in connection with project activities, this episode confirms that I certainly held no monopoly on
power. In this instance I was clearly being met at least halfway by project recipients in pursuit of their own goals. Moreover, women and men
were equally involved in these political maneuvers. The implicit message
was that no pattern of self-sacrificing behavior on the part of men or
women could safely be assumed.

③)   Finally, the Alkalikunda case gave me my first indication that wom-
en's garden projects were especially sensitive from the standpoint of
intra-community relations. As it turned out, of the dozen women's garden projects I worked on over the course of my two years with SCF, virtually all were hampered in some way by conflicts between men and
women. Most problems centered around the question of labor contributions to well-digging projects, but other difficulties surfaced as well.
In one case, a male landholder would not allow the women's group to
erect a permanent fence around their garden for fear of losing access to
the site himself during the rainy season, when he used it to grow groundnuts and millet; in a second, the men's group selected a project site that
most of the women in the garden group opposed (a fact that only came
to light after several project wells had already been dug); in a third site
in a community adjacent to the SCF project area, a delegation of men
successfully lobbied the town chief to ban gardening altogether because
they felt women were neglecting their domestic duties in favor of their
vegetable plots (Schroeder and Watts 1991).

    Part of the reason for these difficulties, I now realize in retrospect, had
to do with the fact that many NGOs working in rural Gambia in the
1980s had prioritized women's projects over men's in response to the
Women in Development (WID) mandate proclaimed by the United Nations (see chapter 1). While SCF's program, for example, was divided
into several "sectors" (child / youth, social development, food production / water supply, health, and credit), each linked in some way to child
survival and general community welfare, individual projects were often
centered on women. The staff of SCF's social development sector had
some of its greatest successes running numeracy / literacy training ses-
sions for adult women. The health program produced dramatic reductions in infant and maternal mortality figures and significantly broad-
ened the range of reproductive choices available to rural women. The
agency's economic development sector ran a revolving credit program
that provided numerous small loans to women while achieving nearly
100 percent repayment rates. And in terms of both levels of participa-

See positive
changes!
women
so keep
doing

tion and favorable impact on rural livelihoods, my own work in the food production sector met with its greatest success through projects involving market gardens managed exclusively by women. In short, SCF's project coordinators often found working with women's groups more productive than working with men. Moreover, they saw work with women as being more central to the agency's mission of improving child welfare. Thus, in 1991, after several years of gradually refining the agency's focus, the country field office director concluded that SCF had become "primarily a women's program." As I hope to demonstrate in this book, the heavy concentration of NGO efforts in program areas favoring women during this period bred resentment on the part of many male residents of the communities the agencies served, and this gave rise to political ecological dynamics that had profound implications for the viability of both women's horticulture and the environmental stabilization projects that soon followed.

THE RESEARCH PROGRAM

The market garden sector received broad support from several different development agencies in The Gambia in the mid-1980s. NGOs, voluntary organizations, and mission groups were heavily invested in garden promotion, often in keeping with policies directed at promoting WID goals. At the same time, however, gardens were often disparaged by larger donors. The United States Agency for International Development (USAID) and what was then known as the European Economic Community (EEC), for example, had their sights set on expanding fresh fruit and vegetable exports to Europe. In the eyes of consultants hired by these agencies, the key to success in the horticultural sector was a group of large-scale commercial vegetable and fruit growers with farms located in the vicinity of Banjul. The primary objective for Gambian horticultural development was therefore the establishment of a "cold chain" to facilitate refrigerated storage of fresh produce and make exports more viable. By contrast, these same experts portrayed rural women's gardens as poorly managed and beset by marketing difficulties, a characterization that had considerable influence in restricting the Department of Agriculture's extension efforts on the women's behalf.

With such priorities in force, there was a paucity of even the most basic information pertaining to women's horticultural production. The broad scope of NGO and volunteer programs notwithstanding, only a handful of studies had been undertaken that explored in any substantive way the systems of production gardeners had developed. Contrary to the impression left by development experts, women's gardens were often quite complex, involving dozens of different crops that met subsistence needs as well as market demand. While the successes of the rural market garden sector were geographically uneven, residents of many North Bank communities became heavily dependent on women's garden incomes. Yet the seasonal pattern, scope, and social significance of the income women earned from vegetable and fruit sales was largely ignored.

Filling these research gaps became the focus of a dissertation project I designed while completing requirements for a Ph.D. in Geography from the University of California–Berkeley. Formal fieldwork for the project was conducted in two phases in 1989 and 1991. I also made a postdoctoral research trip to The Gambia during the summer of 1995. My initial seven-week reconnaissance trip to the North Bank, in July and August 1989, included visits to 18 different communal garden sites and interviews with 127 individuals, including female vegetable farmers, male garden landholders, extension agents, government officials, and agricultural researchers. Scheduled during the seasonal rains, when women typically abandon irrigated gardening in favor of rain-fed rice production, this trip was an opportunity to observe the extent of vegetable production carried out during the off-season and thereby develop a keener sense of the scope of the garden boom and its impact on rural livelihoods. While brief, the visit confirmed several impressions I had formed of petty vegetable commodity production while working on a consulting contract in The Gambia in 1983, and during my two-year residence on the North Bank from 1986 to 1988.

First, I was able to establish clearly that the changes in garden practices dated from the early 1970s. This suggested that they were in some way part of a broad ecological shift that accompanied the gradual decline in rainfall in the region during this period. Second, I confirmed that women in many North Bank villages were in fact very actively engaged in gardening during the rainy season. Given that these activities took place at some cost to their rice-growing responsibilities, the emergence of rainy season gardening indicated the deepening intensification of garden practices. Third, I established that women's cash crop incomes from gardens were substantial and that they permitted women to take on a range of new social responsibilities, most notably a major share of the responsibility for supplementary grain purchases for their households. Finally, I discovered that the gardens had generated disputes—between men as husbands and landholders and women as wives and gardeners—

in several key domains: (1) gardens had become the center of an acute spatial conflict between competing uses of low-lying land resources involving vegetable, rice, and fruit cultivation and livestock grazing, each of which was gendered in very particular ways; (2) the intensification of horticultural production had generated a time-consuming labor regime that kept women away from home for long hours; this produced great resentment on the part of vegetable growers' husbands during the early stages of the garden boom because it cut into the time women allocated to their domestic routines; and (3) the incomes generated by gardeners had prompted considerable jealousy on the part of the women's husbands, both at the household level, where garden incomes frequently exceeded incomes male farmers earned from sales of groundnuts (peanuts), the country's main cash crop, and at the broader community level, where women's groups attracted a disproportionate share of attention from NGOs and other development agencies.

These preliminary findings helped shape the second phase of my research during a ten-month stay from February to November 1991. The principal site of this research was the town of Kerewan, the North Bank divisional headquarters, where I had resided previously during my employment with SCF. This was a site where women gardeners were extremely active but one where I had had no direct involvement in my professional capacity as SCF food production sector coordinator. (While SCF's project sites did not include Kerewan in the period 1986–1988, Kerewan had been incorporated into the agency's program area by the time I did my research in 1991.) I was thus able to build my research plans around a set of long-standing personal contacts with community residents who were friends and neighbors for two years prior to the research, without inviting a direct conflict of interest between my research objectives and my prior engagement on behalf of SCF.

Three systematic surveys consumed most of my time and energy during the first several months of the research. The first, carried out over an eight-week period from February to April, involved a demographic and economic census of 700 domestic units (Mandinka: *dabadaalu*) in 240 residential compounds *(kordaalu)*.[2] Among other findings, this exercise indicated that, in a community with 2,500 residents, roughly 540 women were active vegetable gardeners at the time of the survey. In a second major systematic effort, my research assistants and I selected a sample of 100 gardeners in order to gather yield and marketing data on a weekly basis for a four-month period extending from early February to mid-June.[3] The sample, which consisted of 19 percent of all women gardening in Kerewan, included growers whose plots were located in seven of Kerewan's twelve major garden perimeters then in operation (several other fenced enclosures have been added since). Seventy of the town's *kordaalu* were represented, including three compounds in the top economic census category, fifty-two from the middle economic category, and fifteen from the bottom category, a distribution that was roughly representative of the town as a whole.[4] The women surveyed ranged from fifteen to well over sixty years of age. Twenty-four of the women selected were either unmarried or de facto heads of their own households, i.e., they were widows, their husbands were not co-resident, or they had assumed full financial responsibility for household finances due to their husbands' advanced age or infirmity. Seventy-six male groundnut growers who were married or otherwise financially linked (e.g., fathers) with these women were also surveyed to develop some sense of male income-earning capacity and provide a basis for the comparison of male and female cash-crop incomes. Comparative data were also generated for the Upper Niumi District village of Lameng (sometimes spelled "Lamin") and the Upper Baddibu District villages of Illiasa and Jumansari Baa. A third structured data collection exercise focused on production practices in the gardens. Each woman in the garden income sample[5] was asked to participate in a detailed survey on land tenure, well construction, tree planting, cropping strategies, garden techniques, labor allocation, assistance from male family members, marketing practices, and changes in consumption patterns due to increased garden incomes.

In addition to these approaches, semi-structured interviews were conducted with male landholders and women's garden group leaders on the history of site development, land and tree tenure practices, and the potential for further development programming in each of twelve garden perimeters. These sites, encompassing 19 hectares of land area,[6] 1,370 wells and nearly 4,000 trees, were each mapped, measured, and inventoried as a means of assessing the threat posed by tree crops to garden enterprises. Several disputes over production dynamics and market issues were documented via oral histories and consultation of written reports and archival records. These included a demonstration of several hundred women protesting threatened withdrawal of land use rights in 1984, a market boycott in 1989, the alleged theft of fencing materials by a landholder in 1990, and the displacement of an existing garden by a Norwegian-funded orchard project under way during the research period in 1991.

I also gathered documentary and oral history evidence pertaining to the horticultural policies and practices of state-sponsored and non-governmental organizations involved with horticultural projects. Some of this evidence was generated in formal interviews, and some in the context of over a dozen research briefings I gave to interested agencies. Near the end of the research, I also organized a day-long national workshop to debate my findings and the future of horticultural programming in The Gambia. This session was attended by representatives of the Gambian Government (Departments of Agricultural Research, Services and Planning; and the National Women's Bureau), large donors (United Nations Development Programme, United States Agency for International Development, European Development Fund, World Bank), non-governmental organizations (Save the Children Federation, Action Aid, Methodist Mission Agricultural Program), and voluntary agencies (U.S. Peace Corps, Voluntary Services Overseas [UK]) active in horticultural programming in The Gambia. It served the dual purpose of disseminating the results of the research and eliciting reactions to my findings. In addition to the groups formally represented at the national workshop (many of whom received their own individual briefings), I also provided verbal and written briefings and programmatic suggestions to representatives of the United Nations Food and Agricultural Organization's (FAO) fertilizer project, researchers from the University of Wisconsin Land Tenure Center, and the Gambian-German Forestry Project.

Finally, in June and July of 1995, I made a seven-week followup visit to The Gambia to re-interview the market gardeners in my original research sample. I met with several development agents to discuss shifts that had occurred during the ensuing four years and physically inspected each of the garden sites to assess land use changes.

MALE GENDER RESEARCH(ERS)

The ways in which gender has been invoked in methodological discussions pertaining to social science research on development has changed substantially over the past twenty years. Feminist critics in the late 1970s and early 1980s highlighted the fact that most research on agriculture had up to that point been performed by men and that this gender imbalance within the research community (and the corresponding development bureaucracies) had rendered the economic contributions of women all but invisible. While acknowledging that some of these same blind spots persist in contemporary research practice, recent dis-

cussions have developed a more nuanced understanding of the concept
of researcher identity and its implications for research interactions.
Within the context of these discussions, scholars have been interested in
exploring the theory and practice of "cross-gender" research. There
have been several accounts, for example, written by female researchers
of their attempts to "cross" gender boundaries and conduct ethnographic research among groups principally comprised of male research
informants (e.g., Wolf 1996).[7] Most of these women stress that they
were in some sense "honorary males." That is, their racial, class, or cultural status superseded their gender identities in the minds of their informants, and they were accordingly included in social situations from
which other women in the societies they studied were typically barred.
Thus, the assumption of a cross-gender identity gave women researchers
access to data they would otherwise have been denied.

The research I present in this book was the product of a "crossing"
in the "opposite" direction insofar as both my principal research assistant and I were men and our research subjects were mostly women (cf.
Gregory 1984).[8] Like the women researchers who were treated as honorary males, my assistant and I were in general very warmly received by
our female informants, and this reception warrants further explanation.
In part, the fact that women granted us privileged access to information
pertaining to their garden-based livelihoods can be explained by our individual and collective interpersonal skills and attitudes. I possessed reasonably strong Mandinka language skills and had already lived in the
area for an extended period before I began the research. I was thus in a
position to build the research on intimate personal relationships and
had enough facility with the language to convey proper respect, com-
municate a full range of emotions (sadness, surprise, joy, anger, concern—the basic and ineluctable elements of empathy), and deploy a
sense of humor. These kinds of connections were simply indispensable
for accomplishing what we did. I also had the good sense to rely heavily on my assistant when seeking a compass to guide the research program. As a junior elder and local resident, my research assistant was old
enough to have earned his peers' respect, and by virtue of a remarkably
gentle disposition he was extremely well liked by people of all ages in his
community. Many doors were open to us simply due to the force of his
warm and easy-going personality.

We both worked hard at meeting an unspoken burden of proof
through hundreds of hours of participant observation in the gardens,
most notably in connection with the weekly collection of income data.

This approach is, of course, the very hallmark of intensive ethnographic research, and our efforts were thus not unique in this regard. It was in this connection, however, that our individual and collective identities may have mattered most. I was a well-educated white expatriate who had once worked for an NGO in the area, and my assistant was a member of one of the town's founding lineages, a forestry department employee and a small-scale landholder in his own right. The fact that either of us was interested enough to inquire about gardening at all when others before us had ridiculed or ignored the garden sector went a long way, I believe, toward winning acceptance for our efforts. Moreover, our privileged status meant that we might potentially make ourselves *useful* to the women. These are factors that seemed to play a role, for example, in the Alkalikunda case I recounted above. The women who felt wronged by the head of the men's group who attempted to use my authority against them resisted the temptation to jump to conclusions about my own role in the case. They could have easily assumed that I was abusing the privileges of my power and held a grudge, but instead they brought the matter to me directly. This underscores the fact that the group felt they had something to gain from their interactions with me and shows that, to a certain degree, our interactions were based on the assumption of mutual respect. I believe the research relationships my assistant and I developed in Kerewan were forged on a similar basis.

None of the foregoing should be construed to mean that I feel our gender status played no role in our research. To the contrary, the fact that both my assistant and I were male researchers studying gender relations with a primary focus on women in a rural and heavily Muslim community clearly constrained our choice of methods and limited the scope of the research project. In terms of the practicalities of the research, neither of us was able to interview any of our female research subjects at night. To do so would have invited suspicion of sexual impropriety. Moreover, as we initiated contacts with research subjects, explained the nature of our research, and ascertained their willingness and consent to participate in the study, we were also obliged to meet and discuss the project with the women's husbands, asking the men's permission to talk with their wives. In order to allay suspicion further, we conducted many interviews "publicly," i.e., in plain view on outdoor verandahs or with the doors to sitting rooms open. While the content of these public interviews was never of a very sensitive nature, women may nonetheless have felt constrained against speaking freely under these circumstances. As the research wore on and the novelty of my presence wore off, opportunities for more confidential and potentially sensitive discussions regarding land use or marital politics emerged, and both my informants and I took advantage of them to exchange key information. Interviews conducted in the gardens (located a kilometer or more outside the town proper—see map 3) and away from family compounds were especially useful in this regard.

Ironically, the greatest resistance I encountered in making the status shift my project entailed actually came from my male research subjects. When I arrived in Kerewan for the principal phase of this research in late January 1991, women gardeners were about to begin harvesting the year's vegetable crop. This meant that my immediate priority was to choose a research sample and quickly embark on a yield and income survey. Thus, for several weeks, my research assistant and I virtually ignored the men in the community, and this bred resentment that resurfaced in later stages of the research when we made an explicit attempt to survey male farmers in the area. For example, I was told point blank by one man that my research would never "ripen" (Mandinka: *moo*), or come to fruition, because I was only talking to women and not to men. He argued—quite rightly—that I could not possibly compile a complete picture of the Mandinka agricultural system, much less learn the language properly, if I failed to incorporate men more directly in my research. I could only remain a *toubaab*, a Mandinka / Wolof term that refers to white foreigners in its most general sense, but also to educated elites of any racial or ethnic background who insist on maintaining their privileged status in their interactions with lower status groups. Here the nature of my "gender crossing" was invoked in a very different way. I was a male researcher, and this equated in my informant's mind to a particular affinity for the concerns of men in Mandinka society. When I failed to display evidence of that connection, and maintained instead a persistent focus on women farmers, this called into question the credibility of my research results.

In the latter stages of the research project, one of the husbands of the women in my principal research sample simply refused to be interviewed concerning his own agricultural practices. He complained churlishly that we had only been interested in talking to his wife for several months, and he was not about to begin cooperating with us at such a late date. In effect, with this complaint, this informant seemed to indicate that since I had chosen not to "cross" the racial / class / status divides to investigate male production systems early in my research, I would not now be allowed this crossing. In retrospect, I can also see how

my fairly exclusive early focus on the gardens mirrored the NGO focus *normative* / *did not* / *realize* [handwritten marginalia]
on development projects favoring women. The resentment and occa-
sional resistance men put up against my project is understandable in this
light. Like the garden projects themselves, at a symbolic level, my research focus represented a loss of male power and prestige.

In sum, my gender identity was clearly extremely important to the
outcome of my study, but in very particular ways. What seemed to matter most was precisely how I chose to act with each set of informants and
how my male identity intersected with my race, class, and other status
markers. Moreover, my gender identity was in some ways less an issue
with women informants than I might have expected. If anything, it was
more important in my interactions with men of all ages, who assessed
my masculinity and the degree of affinity I demonstrated with their own
decidedly "male" concerns, and sometimes found them wanting.

THE ARGUMENT

This book has two parts. The first focuses on the emergence of market
gardening as a lucrative livelihood strategy for rural Mandinka women
on The Gambia's North Bank, and the second outlines steps leading
to the introduction of agroforestry practices by men on low-lying lands
which eventually threatened the gardens through shade canopy closure,
the dispossession of land rights, and the redirection of development
benefits. The case study is thus centered on a conflict between two ostensibly "progressive" development objectives that emerged on the North
Bank in the 1980s, one focused on gender equity and the other dedicated
to environmental stability.

Chapter 1 opens with a brief description of the garden boom as it ma-
terialized in the North Bank community of Kerewan. After a review of
a range of theories connecting gender, environment, and development,
I argue that parallel naturalisms embedded in theories of material altruism and ecofeminism are mutually reinforcing. I maintain that these
ideas merged in the 1980s and provided a powerful ideological rationale
for designing development interventions to benefit women. The (re)gen-
dering of development theory and practice coincided with a series of
droughts that virtually spanned the African continent. The sudden influx of large sums of development capital generated to support drought
relief, food security, and environmental initiatives dovetailed neatly with
WID programs and provided the material basis for thousands of projects explicitly focused on women.

Chapter 2 shows how this pattern of gender-sensitive investments served to underwrite the boom in Gambian women's market gardening. After a brief introduction devoted to the history and ethnography of Mandinka agricultural practices on the North Bank, I review North Bank residents' interpretations of the origins of the garden boom. I show how these ideas were roughly split along gender lines, with women showing greater interest in production-related factors, such as climate change and donor contributions, and men emphasizing new consumption patterns growing out of the deepening commercialization of the North Bank economy.

Chapter 3 explores the intra-household budgetary implications of the emergence of a female cash-crop system. I provide data indicating that women's incomes from garden plots often outstripped their husbands' incomes from groundnut production and show how men and women responded to this unprecedented reversal of fortunes in the course of household-level budgetary negotiations. The upshot was a new conjugal contract that left women considerable social mobility and freed men from many of their family financial obligations. I argue that the effect of these negotiations on the garden boom was to exert unrelenting social and economic pressure on women to intensify vegetable production in order to continue securing cash for household needs.

Chapter 4 lays out the social relations of garden production that grew out of the household-level budget negotiations. I show how women carefully calibrated their garden-based production regimes so as to minimize the degree of social disruption their gardening endeavors entailed. Successfully meeting this goal meant forging secure market linkages, overcoming seasonally variable irrigation, crop protection, and disease problems, and carefully integrating domestic duties with garden work tasks.

Chapter 5 takes up the question of land tenure. I demonstrate how women gardeners were able to expand usufruct rights to garden plots to good success for the better part of a decade through tree planting, surreptitious land transfers, and strategic alliances with WID-oriented development agencies. These gains were threatened and in many cases reversed, however, as male landholders began to sense that their landholding rights were eroding. Working in concert with extension agents engaged in promotional efforts directed at agroforestry, several landholders in Kerewan began developing fruit orchards directly on top of the garden plots as a means of reclaiming the land resources in question and redirecting development aid toward personal economic objectives.

Chapter 6 traces the rationale for this rather abrupt shift in aid priorities to the emergence of a new discourse concerning connections between women and their environments. While several Gambian planning documents seemed to bear distinct marks of the influence of ecofeminist and feminist environmentalist thinking, the actual policies they espoused were inconsistent with critical gender perspectives. I describe how the ensuing changes in development practices affected several specific garden perimeters in the Kerewan area and attempt to explain the motivations of the various actors engaged in the resulting land use and labor disputes. I conclude by more precisely identifying the orchard projects as "successional" agroforestry systems premised on commodity production. I argue that such systems almost inevitably entail social and ecological contradictions that undermine many of the projects' best intentions.

Chapter 7 begins with discussion of a debate that has grown up around gender and environmental reforms that have been embraced by the World Bank and other major donors. While some analysts see these policy shifts as progressive developments, others condemn them as forms of co-optation. I underscore the highly ambiguous political character of many gender and environmental interventions undertaken by development agencies and highlight two directions for further research, the first centered on the donors' attempts to co-opt critical ideas, and the second directed at the need to sharpen critiques so as to ensure that they are not used against the very people they were intended to benefit. Finally, I note that women gardeners have made tremendous gains over the course of the past two decades in The Gambia. I argue that these gains should not obscure the fact that recent environmental interventions have worked to their disadvantage, however, and highlight the need to keep questions of power and justice central to any assessment of the political ecology of the market garden districts, particularly as they find expression in the control over female labor.

AUDIENCE

The book is intended for textbook and other scholarly use in the fields of critical human geography, gender studies, development studies, African studies, and forestry. It also speaks directly to the large community of development practitioners and activists interested in the intersection of program objectives relating to gender, the environment, and economic development. In geography, the study marks a contribution to the burgeoning field of political ecology (Blaikie and Brookfield 1987; Bryant 1992, 1997; Bryant and Bailey 1997; Peet and Watts 1996) and most particularly to the body of work centered on the critique of "globalized" environmental interventions; (*The Ecologist* 1993; Sachs 1993; Schroeder and Neumann 1995; Taylor and Buttel 1992). In Africa, this literature has centered on the twin problems of ecological dearth and diversity (Schroeder forthcoming). Whereas in many parts of the continent questions of protecting and preserving biodiversity are paramount (Adams and McShane 1992; Anderson and Grove 1987; Bonner 1993; IIED 1994; Neumann 1995, 1996), The Gambia is situated on the fringe of the Sahel, an area subject to periodic droughts and secular environmental degradation. The principal task facing Gambian natural resource managers has accordingly involved *producing* biodiversity through a program of environmental reconstruction (tree planting and soil and water conservation). Never a simple proposition (cf. Blaikie 1985; Blaikie and Brookfield 1987), this goal was especially difficult to achieve under the conditions of tight fiscal constraint prevailing in The Gambia in the 1980s and early 1990s. A key question faced by Gambian administrators, therefore, and one which infused the low-lying ecologies where Mandinka women's market gardens were located with political economic significance, was how the state and the various development agencies involved in natural resource management could mobilize the labor resources of rural peasants in the service of environmental restoration goals.

The fact that the state and development donors turned increasingly to women as a source of that labor in The Gambia connects the case to gender studies. There are three lines of inquiry that make this study a somewhat unique contribution to this field. First, the study provides a retrospective look at the impact of WID programs in The Gambia some fifteen to twenty years after they were implemented. The detailed historical perspective on WID interventions is itself somewhat unusual, as is the portion of the text devoted to an analysis of their impact on men, who rarely figure in analyses of the WID years. Second, the central narrative of the book describes the considerable success women market gardeners achieved in securing land rights and raising profits from vegetable sales in the face of considerable structural constraints. As I explain in greater detail below, the extensive literature on household relations in Africa (see reviews in Davison 1988; Guyer and Peters 1987; Hodgson and McCurdy forthcoming; Leach 1994; Moock 1986; Parpart 1989) reveals that virtually all social and economic decisions made at the household level in Africa involve careful negotiation between men and women, often producing conflict. Few of these studies, however, depict women entering marital negotiations from positions of comparative economic strength, or show how they wield such political-economic leverage in relationships with men in different structural positions in their communities (but see Mikell 1997). Third, the image of relatively strong and resourceful Gambian women stands in particularly sharp contrast to the hapless victims of environmental decline who figure so prominently in much of the gender, environment, and development literature. This book makes a significant contribution to the gender and environment literature by showing how the debate over some ecofeminist claims that women possess "natural" connections to their environments (Agarwal 1992; Braidotti et al. 1994; Jackson 1993; Leach 1994; Leach et al. 1995; Plumwood 1993; Rocheleau et al. 1996) filtered through the "development apparatus" (Ferguson 1990) in the 1980s and took shape in particular forms of development interventions. Specifically, I demonstrate how the notion of women's "special" status as environmental managers was invoked to justify the use of their unpaid labor in agroforestry projects.

The succession of approaches to women and gender alluded to above speaks to a phenomenon that should be quite familiar to students of development studies, namely, the somewhat fickle nature of development interventions (Escobar 1995; Ferguson 1990; Leach and Mearns 1996; Peet and Watts 1996; Roe 1998; Sachs 1992). At every historical juncture, new development ideologies or sets of political economic circumstances have been invoked by development experts to justify shifts in the ebb and flow of development largesse. The Gambian case illustrates how the opportunities and constraints created by such shifts set in motion parallel moves by specific groups within local polities to reposition themselves and their livelihood strategies on the changed political-economic landscape. The particular pattern of shifts that took place in The Gambia in the 1980s and early 1990s was itself somewhat distinctive, insofar as both waves of intervention, the first focused on gender equity and the second on environmental stabilization, were developed in response to critiques of prevailing development practices by gender and environmental activists. The book highlights the peculiar political dilemmas that ensue when seemingly progressive critiques of development are effectively *co-opted* by donors, stripped of their critical political content, and redeployed to purposes that counter their original intentions.

Finally, the case study makes a compelling addition to the corpus of

studies devoted to the theory and practice of social forestry. In this regard, parallels to the vigorous debate among Indian scholars over connections between gender and forest management in the Chipko movement (Guha 1989; Rangan 1993, 1996; Shiva 1988), and the growing
set of studies exploring the theory and practice of agroforestry (Bryant
1994; Cline-Cole 1997; Dove 1990; Fried forthcoming; Hecht et al.
1988; Leach 1994; Michon et al. forthcoming; Peluso 1992, 1996;
Rocheleau 1988; Rocheleau and Ross 1995; Suryanata 1994; Thomas-
Slayter and Rocheleau, 1995) are especially relevant. Regarding the
latter, I establish a crucial distinction between multidimensional agroforestry systems designed to embrace a diverse array of livelihood strategies and successional systems meant to preserve the relatively powerful
positions of tree cultivators against rival claimants.

Acknowledgments

This research project was generously funded. My primary support for dissertation field work and write-up came from four sources: the Fulbright-Hays Doctoral Dissertation Research Award, the Social Science Research Council / American Council of Learned Societies International Doctoral Research Fellowship for Africa, the National Science Foundation Fellowship in Geography and Regional Science, and the Rocca Memorial Scholarship for Advanced African Studies. In addition, the Rockefeller Foundation funded a short pre-dissertation research trip to The Gambia in 1989, the Rutgers University Research Council supported a return trip in 1995, and the Rutgers University Center for the Critical Analysis of Contemporary Culture supplied a year-long writing fellowship in 1995–96. I am especially grateful to these institutions for their financial assistance.

Numerous colleagues have read all or parts of the manuscript at different stages of its development. I would especially like to thank the following for critical comments, advice, and encouragement: Michael Watts, Louise Fortmann, Dick Walker, Gillian Hart, Dorothy Hodgson, Rod Neumann, Neil Smith, Jane Guyer, Susan Geiger, Marjorie Mbilinyi, David Gamble, Robert Harms, Donald Moore, Sheryl McCurdy, Jesse Ribot, Susanna Hecht, Elon Gilbert, Josh Posner, Oliver Coomes, Don Krueckeberg, Maria Espinosa, Andy Stewart, Colleen O'Neill, Paige West, Judith Mayer, Vinay Gidwani, Tad Mutersbaugh,

Saul Halfon, and the anonymous reviewers for the University of California Press.

The research itself was carried out in The Gambia under the joint auspices of the Ministry of Agriculture's Department of Agricultural Research (DAR) and the Gambian field office of Save the Children Federation / USA. I owe special thanks to Dr. M. S. Sompo-Ceesay of the DAR for sponsoring the research and hosting the national workshop at which my findings were presented and debated. Among SCF staff members, field office directors, Abou Taal and Diane Nell, and program managers, Yahya Sanyang, Lamin Dibba, and Marc Michaelson, were extremely generous with their time and resources. Borang Danjo, Fatou Banja, Faburama Fofana, and Kevin Patterson became directly involved in data collection efforts at various points in the research. Madi Touray, Trisha Caffrey, Lamin Sanneh, Lamin Camara, Satang Jobarteh, Kawusu Hydara, Lamin Saho, Ebrima Jarju, and Margaret Luck were also great colleagues during my tenure with the agency. All of these friends and coworkers supported me in countless ways during my various trips to the North Bank, and I remain deeply indebted to them.

Among the community of expatriate researchers and development workers I encountered in The Gambia, I owe a major intellectual debt to Judy Carney, whose work on gender and household conflicts in the irrigated rice perimeters along The Gambia's South Bank set the standard for such studies not just among Africanists but within the scholarly community at large. David Gamble was no longer physically present in Kerewan when I first arrived in the mid-1980s, but his work in the community as an economic anthropologist attached to the colonial civil service in the 1940s certainly lived on in local memory. His studies provided important baseline data for assessing the changes accompanying the garden boom. Don Drga and Philip DeCosse were especially generous in sharing documentary materials, and it was Philip who first suggested a national workshop as a way to ensure that the results of my research would be put to good use. John Bruce, Mark Freudenberger, Sarah Norton-Staal, and Cathy Jabara were helpful and cooperative as we explored our mutual research interests. Elon Gilbert and Josh Posner were highly valued critics. My thanks to all of these individuals and to friends, Bala Sillah, Joan Millsap, Dawn Vermilya, Mike Grebeck, Julia Rumbold, Steve Harris, Jim Zinn, Patricia Fialcowitz, and Andrew Watt, for their help and support.

My debts to people in Kerewan are many and varied. I would like to first thank the hundred women and their families whose lives and livelihoods I studied and wrote about here. Early on in the research, my primary research assistant, Maline Hydara, and I developed a shorthand for certain situations we encountered in the course of our inquiries. The Mandinka phrase, *A bee korontoring,* means quite simply, "She's in a hurry." Its implication was "This woman is too busy to talk; let's not bother her." In reality, of course, most of the women we talked to were too busy to talk most of the time, and yet they still agreed to be interviewed and answer our questions. I can only hope that I have done their story justice in return. To Maline, with whom I shared this project from start to finish, I owe my deepest gratitude. He is a man of profound sensibilities and unfathomed good humor and wisdom. Many thanks, Serifo. *Ala maa mbeela jari la.*

In the States, I have first to acknowledge the members of my graduate committee, Michael Watts, Dick Walker, Louise Fortmann, and Gillian Hart. Most graduate students are lucky to have even one person of this caliber as a mentor. I had four. Would that I could live up to all the goodwill and support they showed me in the course of my time at Berkeley. It was a special privilege to have Michael as my principal advisor. He was peerless as an intellectual mentor and role model, extremely supportive, and a warm and vital friend to boot. Cheers, Michael. It was great.

Among my geography colleagues at Berkeley, George Henderson, Susan Craddock, Frank Murphy, Susanne Freidberg, Victoria Randlett, and Tad Mutersbaugh were great friends throughout. I owe special thanks to Allan Pred, Rod Neumann, Brian Page, Eric Hirsch, and Katharyne Mitchell. Allan might cringe at the thought, but it's he who almost single-handedly defined geography for me. Brian and Eric helped demystify the graduate school and grant-writing game at every turn. Katharyne's letters while in the field made me laugh to tears—a true balm for the soul. And Rod's support and incisive intellectual engagement both at Berkeley and since have been absolutely critical to the success of this project and many others. At Rutgers, my faculty colleagues in Geography and African Studies have created a work environment that is both consistently stimulating and largely free of the sorts of tensions that characterize many academic programs. A number of graduate students, including Sandra Baptista, Renaud DePlaen, Marla Emery, Salvatore Engel-DiMauro, Maria Espinosa, Ruth Gilmore, Noriko Ishiyama, John Kasbarian, Karen Nichols, Andy Stewart, Bansuri Taneja, and Paige West, have conspired in different ways to keep my scholarship sharp. Mike Siegel's cartographic skills have been extremely helpful, and Elaine

Gordon and Betty Ann Abbatemarco have provided excellent administrative support. Thanks also to Sandra Baptista for research assistance.

Pride of place here is reserved for long-time friends and family. In The Gambia, Sean O'Leary, Janis Carter, Ted Wittenberger, Cathy DeSanto, and Barbara Kah pulled me back from more than one brink. In the States, Paul Lopes bore the full brunt of that task, along with Tidiane Ngaido, Daniel Schneider, Marc Rosenthal, Julie Derwinski, Jenny Kirst, Dave Rez, Nancy Shillis, and Ned Hamilton. My brothers and sisters have been completely unwavering in their love and support through some occasionally difficult times. To Tim and Jan, Lal and Kirk, and Tom and Jenny my love and thanks. My surrogate family at Rutgers includes Barbara Cooper, Richard Miller, Cindi Katz, and Neil Smith. Barbara and Richard are, among other things, paragons of parental virtue, and they have taught me by example how to strike the ever elusive balance between work and family responsibilities. Cindi and Neil are simply awe-inspiring in their capacity to do it all, do it well, and keep having fun. Their generosity of heart and mind expresses itself in ways I could not have imagined before they welcomed me into their lives. My son, Luke, graced the world just as the final draft of this book was being readied to go to press. His sweet smile and delighted squeals of laughter have been all-sustaining. Finally, my wife, Dorothy Hodgson, has been a wonderful companion and source of intellectual inspiration throughout the writing of this book. She is the true scholar of the family, and my admiration and respect for her know no bounds. All my love, Do, I could not have done this without you.

Highland Park, New Jersey, June 24, 1998

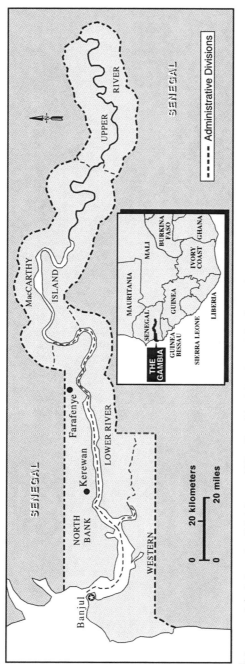

Map 1.   Administrative Divisions of The Gambia. Map by Michael Siegel, Rutgers Cartography.

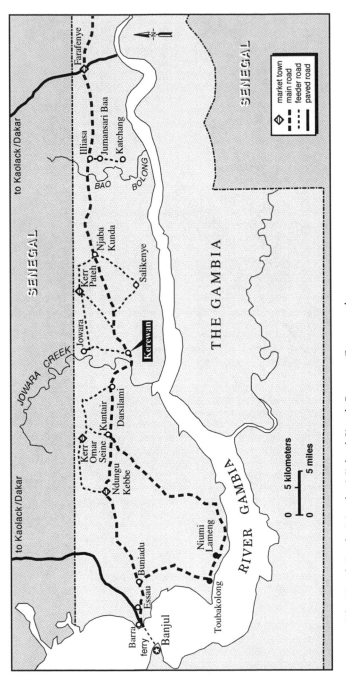

Map 2. The North Bank. Map by Michael Siegel, Rutgers Cartography.

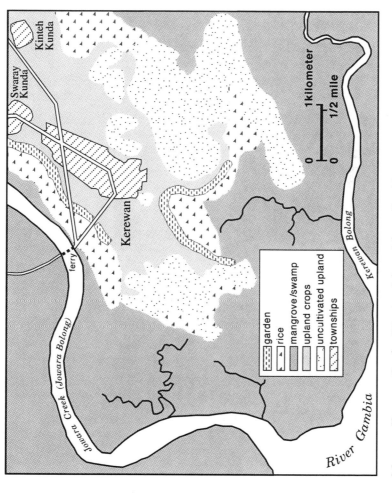

Map 3. Kerewan Garden District. Adapted from Gamble 1955.

Introduction

Some sixty kilometers upriver along the North Bank of The River Gambia lies the Mandinka-speaking community of Kerewan (ke'-re-wan). The dusty headquarters of The Gambia's North Bank Division is located on a low rise overlooking rice and mangrove swamps and a ferry transport depot that facilitates motor vehicle transport across Jowara Creek (Jowara Bolong), one of the River Gambia's principal tributaries. Since the Kerewan area was dominated by opposition political parties throughout the nearly thirty-year reign of The Gambia's first president, Al-Haji Sir Dawda Jawara (1965–1994),[1] it became something of a developmental backwater. Before 1990, Kerewan town had no electricity or running water beyond a few public standpipes. For a community of 2,500 residents, there were no restaurants and only a poorly stocked market that lacked fresh meat. Indeed, from the standpoint of the civil servants assigned to the North Bank Division (see map 1), Kerewan was considered a hardship post. Mandinka speakers sarcastically referred to the divisional seat as "Kaira-wan," a place where "peace" (Mandinka: *kaira*) reigned to the point of overbearing stagnation. Neighboring Wolof speakers, meanwhile, disparaged the community by dubbing it "Kerr Waaru"—"the place of frustration."

Kerewan's reputation was only partially deserved, however, for the community was actually the center of a great deal of productive economic activity. Over two decades beginning in the mid-1970s, the town's women transformed the surrounding lowlands into one of the key sites of a lucrative, female-controlled, cash-crop market garden sector. A visitor to Kerewan as recently as 1980, when I made my first trip to The Gambia, would have found that vegetable production on the swamp fringes ringing Kerewan on three sides was decidedly small-scale. Most gardeners, virtually all of whom were women,[2] worked single plots that were individually fenced with local thorn bushes or woven mats. Outside assistance in obtaining tools, fences, and wells was minimal. Seed suppliers were not yet operating on a significant scale, and petty commodity production was largely confined to tomatoes, chili peppers, and onions. The market season, accordingly, stretched only a few weeks, and sales outlets were all but nonexistent. Most Kerewan produce was sold directly to end users in the nearby Jokadu District by women who transported their fresh vegetables by horse or donkey cart and then toted them door to door on their heads (a marketing strategy known as *kankulaaroo*).

By 1991, when I completed the principal phase of research for this book, large gardens on the outskirts of Kerewan had come to dominate the landscape (map 3). Each morning and evening during the October–June dry season, caravans of women plied the footpaths connecting a dozen different fenced perimeters to the village proper. Over the course of nearly twenty years, the number of women engaged in commercial production rose precipitously from the 30 selected to take part in a pilot onion project in the early 1970s to over 400 registered during an expansion project in 1984, and some 540 recorded in my own 1991 census. The arrival of the first consignments of tools and construction materials donated by developers for fencing and wells in 1978 initiated an expansion period which saw the area under cultivation more than triple in size, growing from 5.0 ha to 16.2 ha in ten years. Between 1987 and 1995, a second wave of enclosures nearly doubled that area again. At least a dozen separate projects were funded by international NGOs, voluntary agencies, and private donors. These donations were used toward the construction of thousands of meters of fence line and roughly twenty concrete-lined irrigation wells. In addition, there were some 1,370 hand-dug wells and nearly 4,000 fruit trees incorporated within Kerewan's garden perimeters. Growers purchased seed, fertilizer, and other inputs directly from an FAO-sponsored dealership in the community and sent truckloads of fresh produce to market outlets located up and down the Gambia-Senegal border, which thrived on the vegetable trade. In sum, the Kerewan area developed over two decades into one of the most intensive vegetable-producing enclaves in the country.

obstacles to garden wide + formidable

Exactly how all of this was possible is one of the questions I attempt to answer in this book. For the obstacles to the garden boom were formidable. At the point of production in low-lying areas where most communal garden sites were located, gardens were forced to compete with rice plots, fruit orchards, and foraging herds of cattle and small livestock for space and access to water. *agric. compet.* The specific points of contact and potential conflict between each of these production systems followed a certain temporal rhythm determined by the seasonal growth pattern of individual crops, interseasonal variation of cropping strategies and range management techniques, and successive stages of tree growth. These spatial and temporal dynamics carried with them distinct political ecological implications that were expressed in the "idioms" of property rights and labor claims (Watts 1993). In order to gain access to land suitable for gardens, protect it from livestock damage, and tap into existing groundwater reserves, women first had to secure usufruct rights from male *need rights in oppressed system* landholders and then leverage funds from developers for fencing materials and well construction. Finally, and most critically, gardeners somehow had to regain control over their own labor in the face of a wide range of competing demands that were often carefully monitored by the women's husbands or senior female relatives.

A second set of dynamics involved working out ways to generate a worthwhile return for garden labor. Once gardeners had solved the production problems imposed by the unforgiving climate conditions along the southern boundary of the Sahel (in large part through backbreaking efforts at hand-irrigating their crops), they were then faced with negotiating profitable marketing arrangements with truck drivers and vegetable buyers. Moreover, securing the economic benefits of the garden-based livelihood system entailed defending crop rights on the *crop rights* home front. In Mandinka society, the right to sell or otherwise use or dispose of a particular agricultural commodity has varied historically, often by gender and on a crop-by-crop basis (Gamble 1955; Haswell 1975; Weil 1973, 1986). For example, with the notable exception of the large pump- and tidal-irrigated swamps Judith Carney and Michael Watts studied in central Gambia (chap. 2; cf. Carney and Watts 1990, 1991), the rice crop was grown by women but was ultimately destined for joint household consumption and could not be sold. By contrast, most fresh vegetables grown on the North Bank were clearly destined for market, and these sales generated revenues that were considered private income. As gardeners' incomes grew, negotiations over household budgetary obligations became a key site of intra-household struggle.

A third set of problems women encountered involved keeping pace with the sometimes fickle policies and practices of development agencies. In its early stages, the garden boom was heavily supported by voluntary agencies, NGOs, mission groups, and other donor agencies. This assistance included material support in the form of construction supplies, tools, and small loans, technical support for well construction, and, in some cases, legal and political support in the form of advocacy on behalf of women's land rights. Beginning in the mid- to late 1980s, however, the prevailing development paradigm shifted, and many of the agencies that had once supported the garden boom began working at cross-purposes with the women vegetable growers. Under the guise of sustainable development, several of the larger international donors providing direct aid to The Gambia embarked on new initiatives to address environmental problems in the region, and NGOs and voluntary agencies quickly followed suit. In practical terms, this meant that some funding for gardens was withdrawn and that a new set of competing claims emerged with respect to the low-lying lands where the women's gardens were located. Specifically, developers were attracted to the irrigation potential of these areas and sought to tap it for the purposes of establishing woodlots and orchards through agroforestry practices. Moreover, they sometimes justified their efforts to reclaim the lowlands on the grounds that agroforestry practices helped involve women directly in environmental management, this despite the threat tree planting in gardens sometimes posed to the women's horticultural efforts. On the face of it, policy measures that first promoted women's gardens and then sought to convert them to fruit orchards or woodlots seem contradictory. Most orchards and woodlots were in fact controlled by male landholders rather than female gardeners, and their immediate effect was to displace gardens altogether in many cases. This inconsistency can only be explained in light of the shifting ideologies that motivated development interventions in the region at the time these changes took place in the mid-1980s.

THEORIES CONNECTING GENDER, DEVELOPMENT, AND THE ENVIRONMENT

The image most widely used to capture the "plight" of Third World women is that of an African peasant woman toting an improbably large and unwieldy bundle of firewood on her head (fig. 1; see discussion in Braidotti et al. 1994). She may or may not have a young child tied to her

Figure 1.   Gender and environmental management. Hauling firewood: a key icon of the gender, environment, and development
literature.

back, but the image is always meant to convey that she has traveled a great distance to gather her load. As a metaphor, this feminine icon suggests the incredible burdens women shoulder, and the great lengths they go to, to satisfy the multiple and competing demands society and their families place on them. The implication is that women suffer these conditions universally, as a class, and that pure, selfless motives drive them to undertake routinely dull, repetitive, and ultimately thankless tasks. At the same time, the graphic portrayal of firewood collectors is meant to underscore the idea that close connections exist between women and the natural environment. It suggests that women forced to gather wood from the countryside lead a hand-to-mouth existence, where knowledge of the landscape is bred of necessity and deep personal experience, and where the vagaries of climate and ecology have profound and immediate implications for human well-being. Thus, by virtue of their collective lot in a singular division of labor, women mediate the relationship between nature and society, and they feel the brunt of natural forces as a consequence.

Such images convey a stark reality: life for peasant women is often filled with considerable toil and drudgery. Yet if these women suffer a common plight, it resides not in any particular niche in some all-encompassing division of labor but in the countless ways the range and variety of their lived experiences are distorted in the words and images conveyed by outsiders (Parpart 1995; Jackson 1993). The wood-gathering icon represents Third World women as Africans, African women as peasants, and peasant women as a single type. There is no geographical detail at either localized or macropolitical scales that might serve as an explanation for the plight thus portrayed. Moreover, to render such women as beasts of burden, dumb, stolid, unwavering in their support of their families, unstinting in their service of same, is to acquiesce in the notion that they are perpetual victims, steeped in need, and incapable or disinclined to contest their lot creatively.

This tension between images of women as victims and women as autonomous actors traces back to the earliest efforts of developers to promote Women in Development (WID) programs in the Third World. The United Nations–sponsored convocation in Mexico City in 1975 proclaimed an International Decade for Women and initiated efforts within the major development agencies to address a broad agenda of issues deemed especially pertinent to Third World women. The image of the long-suffering African woman figured prominently in these discussions and debates, largely because of the influence of Ester Boserup *women's incentives*

(Boserup 1970). Her widely cited work set forth several related propositions: (1) that women in general, and African women in particular, play key roles in most rural production systems; (2) that women's contributions to the food security of their families are made in the face of numerous obstacles, including limited access to land and labor resources and systematic exclusion from the benefits associated with technological change; and (3) that, therefore, there is considerable scope for increasing food production by women through a combination of financial and technical assistance measures. The basic development strategy that flowed forth from these premises had as its primary objective the achievement of "modernization with equity" (Bandarage 1984; Roberts 1979). In effect, the early WID activists advocated reformist rather than structural changes, calling for more equitable distribution of development benefits rather than substantive changes in the development process itself.

This position came under attack from several directions. Critics of WID objected to both the use of "women" as a universal, undifferentiated category and the unidimensional emphasis on women, per se, as opposed to an exploration of the full scope of gendered social relations. They noted the analytical confusion between production and (social and biological) reproduction and the cultural biases privileging the goals and aspirations of Western women over their Third World counterparts. They were also particularly distressed by the lack of a direct challenge to the existing power structure implied by the blind acceptance of the premises of modernization theory (Bandarage 1984; Braidotti et al. 1994; Kabeer 1994; and Parpart and Marchand 1995 offer excellent reviews of this literature).

Such criticism gave rise to two main alternatives to the "Women in Development" paradigm, each objecting to one of the main terms of reference incorporated in the WID framework. The first, known as "Women and Development" (WAD) was essentially a critique of development practice. Its proponents objected to male-dominant, top-down approaches. In their view, substantive change for women could not result from simply reforming this existing development apparatus. Instead, WAD activists argued in favor of a grassroots program of participatory, women-only projects (Parpart 1995). The second major alternative to WID, known as Gender and Development (GAD), took as its point of departure a critique of the essentialist views of "women" embedded in both WID and WAD approaches. Drawing heavily on the writings of socialist feminists and a group of Third World feminists (Sen and

Grown 1987), the GAD position offered a theory that sought to explain gender relations as power relations that grow out of specific social and political contexts:

> [The GAD] perspective leads to a fundamental re-examination of social structures and institutions, to a rethinking of hierarchical gender relations and ultimately to the loss of power of entrenched elites, which will effect [sic] some women as well as men. It focuses on both the *condition* of women, i.e. their material state in terms of education, access to credit, technology, health status, legal status, etc., and the *position* of women, i.e. the more intangible factors inherent in the social relations of gender and the relations of power between men and women . . . (Rathgeber 1995, 206, emphasis in the original)

The WAD and GAD critiques have had enormous influence on the international debates concerning gender and development politics,[3] yet many of the ideas central to the original WID formulations still hold remarkable sway in the major development agencies. One reason for this persistence is that it took these agencies until the mid-1980s to formally adopt WID as a programmatic emphasis. The United Nations Development Program (UNDP) and the World Bank, arguably two of the most influential multilateral development institutions, established their WID programs only in 1986 and 1987, respectively.[4] A second reason for the persistent essentialism stems from the fact that development efforts directed at women have been premised on the dubious notion that women possess inherent capacities for nurturing and self-sacrifice. These ideas derive their staying power in part from a selective reading of contradictory empirical evidence, and in part from the recurrent tendency to naturalize gendered power structures. The problematic implications of such notions can be illustrated by comparing two sets of development ideologies and practices, one focused on the relationship between women and their families, and the other on the relationship between women and the environment.

MATERNAL ALTRUISM

Analysts have long been influenced by a substantial body of empirical evidence suggesting that women across categories of class, race, ethnicity, and national origin devote a significantly greater proportion of their economic assets and physical and emotional energies toward the tasks of providing and securing their families' well-being than do men. Whitehead coined the phrase "maternal altruism" to refer to the ideology that stipulates that women are, by virtue of their identities as mothers and wives, "naturally" predisposed toward nurturing and self-sacrifice (Whitehead 1981; see also Harris 1981). Patterns of altruistic behavior have been observed among certain groups of women in the areas of food and health care provisioning,[5] and the primary role women play in reproductive labor in many societies has been well documented (Beneria and Sen 1981; Deere and Leon 1987). Such findings have led analysts to generalize broadly regarding women's economic motives and behavior: "Virtually all income in women's hands . . . is devoted to household and family expenditures, reflecting the socially ascribed roles of women in meeting daily welfare needs. Increasing the income of rural women thus may be one of the surest ways of increasing basic family welfare in the countryside" (Jiggins 1994, 200; see also Palmer 1991). Furthermore, these ideas have led development planners in such areas as women's income generation and Maternal and Child Health Care (MCH) projects to proceed on the basis of erroneous assumptions. From this perspective, any income generated for women by whatever means—through craft production or the cultivation of cash crops—will produce benefits for a broad social circle. In terms of health care, practitioners have reasonably linked prenatal and neonatal maternal nutrition and health status with the health status of infants and young children. The assumptions many health care professionals share that women are solely responsible for caregiving is much more problematic, however. Schoonmaker-Freudenberger's (1991) insightful analysis of an integrated health and nutrition project funded by UNICEF (United Nations International Children's Emergency Fund) in rural Senegal the late 1980s offers a case in point. This project failed to meet one of its central goals—the increased consumption of health care services via enhanced women's income generation—partly because planners failed to realize that the health and day care services provided by the project were funded by fathers rather than mothers in almost all cases.

The evidence drawn from the Senegalese case that men play specific "nurturing" roles in many societies is only one type of empirical finding that works against the maternal altruism model. There is also an extensive literature that demonstrates clearly and unequivocally that women in a variety of material circumstances routinely work toward private economic ends that have little to do with meeting the joint welfare needs of their families. Where nurturing models see familial social relations as revolving around a natural core of "maternal altruism," the "new household" literature views them as being governed by an ever-changing "conjugal contract" (Guyer 1984; Harris 1981; Whitehead 1981, 88, 107). The household unit is instead identified as "the site of separable, often competing, interests, rights and responsibilities" (Guyer and Peters 1987, 210). Research has demonstrated that the question of which marriage partner pays for necessities such as food, children's schooling and clothing, health care, and petty daily expenses is determined through constant struggle in many African households (Dwyer and Bruce 1988; Folbre 1986; Guyer 1988; Muntemba 1982; Whitehead, 1981). Notions of maternal altruism are confounded by evidence that shows "husbands and wives lending each other money at rates only slightly less usurious than the prevailing market rate, the payment of wages inside households, wives selling water to husbands in the fields, husbands selling firewood to wives, and wives and husbands selling each other animals that are consumed by the family on feasts and special occasions" (Gladwin and McMillan 1989, 350). Similarly, access to land, water, and other natural resources available to the household is contested by marriage partners, and between and among men and women co-resident in households but of different generations (Davison 1988; Dey 1981; Longhurst 1982; Roberts 1979; Thomas-Slayter and Rocheleau 1995). The allocation and remuneration of intra-household labor is particularly fraught with tension, especially under conditions of rapid political economic change, such as climate perturbations, the introduction of new crop varieties by developers, and market fluctuations for key commodities and inputs (Berry 1984; Carney 1988a, 1988b; Carney and Watts 1990, 1991; Goheen 1996; Jones 1986; Leach 1994; Mackintosh 1989; Madge 1995a; Mbilinyi 1991; Moore and Vaughan 1994; Roberts 1988). All of these processes of negotiation and accommodation hold potential for undermining the assumption that women exhibit altruistic tendencies toward family members.

The new household theorists do not, as a rule, deny the evidence that specific groups of women act in apparently selfless ways to support and protect their families and communities under particular circumstances. They recognize the virtues of such social behavior and endorse it in principle (see discussion in Whitehead 1981). The distinction they maintain, and one I endorse, is that such self-sacrificing behavior is neither inherent nor natural, and that the ascription of altruistic motives exclusively to women has clear political consequences. The assumption of maternal altruism leads to the *expectation* that women will deny themselves and shoulder additional burdens in the interests of family well-being. Development initiatives are sometimes designed accordingly, with women trapped in domestic roles by the very programs intended to benefit them (see illustrations in Stamp 1990).

WOMEN AND THE ENVIRONMENT

Whereas nurturing behavior is widely ascribed to women within the household unit, a parallel "naturalism" (Guyer 1984), a kind of "maternal altruism"-writ-large, often lies behind the association of women with the environment (Haraway 1989; Jackson 1993). The political and philosophical movement known as "ecofeminism," which was formed around the notion that women have a "special" relationship with the environment, has generated a lively debate within the feminist community. Rocheleau et al. (1996) count ecofeminism as one of at least six separate bodies of theory and practice that bring together feminism and environmentalism (see parallel discussions in Plumwood 1993 and Leach et al. 1995). Some of these positions, which include ecofeminism, feminist environmentalism, socialist feminism, feminist poststructuralism, environmentalism, and the authors' own approach, which they call "feminist political ecology," were developed in direct response to the emergence of the ecofeminist discourse. Taken together, the different theories embrace a wide range of political perspectives.

According to Rocheleau et al., ecofeminists "posit a close connection between women and nature based on a shared history of oppression by patriarchal institutions and dominant Western culture.[6] Some ecofeminists attribute this connection to intrinsic biological attributes (an essentialist position), while others see the women / nature affinity as a social construct to be embraced and fostered" (Rocheleau et al. 1996, 3). The distinction between the essentialist ecofeminist position and social ecofeminism has important political implications. As a political principle, the emphasis on the struggle against patriarchal domination places women in a subordinate, defensive position. The more socially oriented position, while still premised on a naturalized connection between women and the environment, constructs the position of women in more positive terms. By virtue of the work they do, women are engaged in a relationship with nature that is characterized by "reciprocity, symbiosis, harmony, mutuality and interrelatedness" (Braidotti et al. 1994, 93). This "closeness" is constituted as a distinct advantage for women because it yields "privileged knowledge." The development community, and most particularly the group of liberal feminists Rocheleau et al. identify simply as "environmentalists,"[7] came to appreciate this dis-

tinction belatedly in the midst of global ecological crisis and made it the
basis of a broad range of development policies. As Braidotti et al. explain, "In the late 1980s . . . [t]he images of women in the South as victims became transformed into images of strength and resourcefulness. In
the wider debate on sustainable development women were increasingly
promoted as 'privileged environmental managers' and depicted as possessing specific skills and knowledge in environmental care" (Braidotti
et al. 1994, 88; see further discussion in Leach et al. 1995). This process
by which the core ecofeminist principles, with all of their problematic
assumptions concerning "natural" connections between women and the
environment,[8] were adopted by development agents and incorporated
into development plans is one of the central concerns of this book.

My own position in these debates is somewhat eclectic, drawing insights from each of the remaining "schools" of thought identified by
Rocheleau et al., but aligning most closely with "feminist political ecology" and "feminist environmentalism."[9] According to these authors,
feminist political ecology:

> begins with the concern of . . . political ecologists who emphasize decision-
> making processes and the social, political and economic context that shapes
> environmental policies and practices. Political ecologists have focused largely
> on the uneven distribution of access to and control over resources on the ba-
> sis of class and ethnicity. . . . Feminist political ecology treats gender as a criti-
> cal variable in shaping resource access and control, interacting with class,
> caste, race, culture, and ethnicity to shape processes of ecological change,
> the struggle of men and women to sustain ecologically viable livelihoods,
> and the prospects of any community for "sustainable development." (Roche-
> leau et al. 1996, 4)

The closely related "feminist environmentalist" position is often identified with the work of Bina Agarwal (1992). In a pivotal essay, Agarwal
argues that by emphasizing natural and spiritual bonds between women
and the environment over the material circumstances that shape those
ties, many ecofeminist writers have simply reproduced limited and ultimately self-defeating notions of femininity. In effect, she contends that
ecofeminists have fallen into many of the same analytical traps as those
who subscribe to uncritical notions of maternal altruism. In addition to
persistent essentialism, and a one-dimensional theory of power that ne-
glects cross-cutting relations of class, race, and ethnicity, the core position of ecofeminism:

> locates the domination of women and of nature almost solely in ideology,
> neglecting the (interrelated) material sources of this dominance (based on

economic advantage and political power). . . . [E]ven in the realm of ideological constructs, it says little . . . about the social, economic, and political
structures within which these constructs are produced and transformed. Nor
does it address the central issues of the means by which certain dominant
groups . . . are able to bring about ideological shifts in their own favor and
how such shifts get entrenched. . . . [T]he ecofeminist argument does not take
into account women's lived material relationship with nature, as opposed to
what others or they themselves might conceive that relationship to be. (Agarwal 1992, 122–123)

The materialist focus Agarwal advocates is one broadly shared by
several major theorists of gender, environment, and development working in Africa (Berry 1984, 1989, 1993; Carney 1988a, 1988b, 1993;
Jackson 1993, 1995; Leach 1991, 1992, 1994; Mackenzie 1990, 1991,
1994; and Rocheleau 1987, 1991, 1995). In a wide range of African
contexts, women are deeply involved in environmental management by
virtue of the simple fact that they do so much of the physical labor in
their respective societies. Drawing water, gathering firewood, cultivating
crops, processing and storing food, and collecting medicines and apply-
ing treatments—these tasks are often the primary, if not sole, responsibility of women (Rodda 1991). By virtue of their disproportionately
heavy workload in these areas, women are often directly disadvantaged
by environmental degradation and decline as the labor required to perform routine tasks increases. By the same token, since women are directly responsible for day-to-day management of many vital resources,
they constitute the key to success in promoting new strategies of environmental rehabilitation and repair. It is their regular contact with the
resource base, born of specific work responsibilities, then, rather than
some sort of natural symbiotic relationship, that has given particular
groups of women privileged knowledge of resources their communities
depend upon.

This analysis of the "women and the environment" question provides
critical insights for sorting through the politics of the recent developmental interventions that invoke either the roles of women or gender as
central organizing principles in environmental management:

[Rural women forest dwellers and cultivators] could . . . be seen as both
victims of the destruction of nature and as repositories of knowledge about
nature, in ways distinct from the men of their class. The former aspect would
provide the gendered impulse for their resistance and the response to environmental destruction. The latter would condition their perceptions and
choices of what should be done. Indeed, on the basis of their experiential understanding and knowledge, they could provide a special perspective on the

processes of environmental regeneration, one that needs to inform our view of alternative approaches to development. (Agarwal 1992, 126–127)

The problem comes, as Agarwal has indicated, when these work routines are stripped of their contextual significance, and when the relationship between women and nature is naturalized and essentialized.

Jackson has noted that an emphasis on the simple fact that women possess key environmental knowledge obscures the difficulties of gaining access to that knowledge and making it the basis of a program of environmental initiatives (Jackson 1993). Furthermore, as Leach argues in her insightful study of the gendered nature of forest politics in Sierra Leone, the observation that women are highly dependent on the environment for meeting their needs, and thus important targets for incorporation into environmental management programs, might well "impl[y] that any outside intervention would be a help, and that women would willingly participate because they have no choice" (Leach 1994, 25). In a similar vein, women as "managers" of the environment very quickly become "assets to be 'harnessed' in resource conservation initiatives" (Leach 1994, 25). In sum, "[The] arguments [of some ecofeminists] that close conceptual links between women and nature can provide the basis for an ecologically sustainable future thus run the risk of, in effect, giving women responsibility for 'saving the environment' without considering whether they have the material resources to do so" (Leach 1994, 34; see also Plumwood 1993).

What the foregoing discussion suggests is that the notions of maternal altruism and a naturalized connection between women and the environment are mutually reinforcing. The role of women as environmental managers derives directly from their role as household managers. Their presumed skills in providing "environmental care" derive from maternalistic impulses generated in the context of familial relations. Assumed links between family and environmental well-being add weight to assumptions that environmental interventions focused on women will bear special dividends for all. In this regard, development approaches to women and the environment simply magnify problems that originated in WID approaches based on notions of altruism and self-denial.

If the different ways of naturalizing gender relations reinforce each other, so, too, do their respective critiques. Where the new household theorists have encouraged us to inspect the internal workings of household dynamics, critical feminist environmentalists and feminist political ecologists have urged us to deconstruct and denaturalize the working re-

lationships that shape the nature of the gender and environment encounter. We are encouraged by the advocates of each of these positions
to look inside the household and throughout the different echelons of
the state and the international development agencies to understand how
connections between particular forms of gender relationship and specific approaches to environmental management have been forged.

THEORY AND PRACTICE:
GENDER, ENVIRONMENT, AND DEVELOPMENT
IN AFRICA IN THE 1980S AND 1990S

The particular constructions of female identity that surfaced in the
household studies and gender and environment debates have helped
shape the policies and practices of development and state agencies
throughout Africa. The remaining chapters of this book will be devoted
to an examination of these policies as they emerged in development interventions along The Gambia River Basin in West Africa. The first half
of the book will explore the phenomenon of the garden boom and its implications for gendered social relations in the Mandinka communities
of The Gambia's North Bank. The second documents the shift in development policies that took place in The Gambia in response to the environmental mandate that emerged in the mid- to late 1980s. Both
sections of the book look critically at the various ideas underpinning
development policies and the ways in which the agencies' actions were
received by men and women in rural communities. Before proceeding to
that discussion, however, it is worth briefly reviewing the broad political economic context in which these interventions occurred.

The overriding material condition affecting life and livelihood in
Africa during the late 1970s and 1980s—the years that WID and gender and the environment surfaced as major development motifs—was
persistent drought. This phenomenon had several dimensions: the actual agroclimatological conditions that limited the ability of rural populations in many parts of the region to reproduce themselves from one
year to the next, the famine conditions that ensued in some areas as
drought-related food shortages were exacerbated by civil war and economic crisis,[10] and the full-scale, crisis-oriented relief and rehabilitation
interventions that swept the region in its aftermath.

The geographical distribution of drought is difficult to map with pre-
cision, given the remarkable variability of rainfall that occurs across spa-

tial and temporal scales on the continent. Nicholson's (1985, 1993) research defines a Sahelian zone in West Africa as an area loosely bounded by Mauritania and Guinea in the west and Chad and Northern Cameroon in the east. This region experienced generalized drought conditions from 1968 to 1973,[11] a particularly bad year in 1977, dramatic shortfalls from 1982 to 1984, generally poor rains from 1985 to 1988, and finally a brief upturn as rainfall levels in 1989 and 1990 returned to the long-term average. In Southern Africa, the Kalahari region experienced above average rainfall in the 1970s and a dry cycle similar to the Sahel in the 1980s, with Zambia and Botswana experiencing what climatologists called "the worst climatic disaster in the recorded history of the subcontinent" (Robbins 1983, 56). East Africa, defined as Kenya, Tanzania, Rwanda, Burundi, and Uganda, was "abnormally wet" over an extended period beginning in 1977 and continuing throughout the 1980s. These conditions stood in sharp contrast to the Horn of Africa to the north, where devastating droughts and horrific famine compounded by civil war laid siege to parts of Ethiopia, Somalia, and the Sudan for the better part of a decade.

The material conditions produced, in part, by drought included sharp reductions in food production, devastatingly high mortality rates in livestock herds, deforestation and denudation of landscapes, massive relocation of rural populations, and the outbreak of disease epidemics that contributed to hundreds of thousands of human deaths. Residents of rural communities were forced to resort to a range of increasingly desperate measures to help them cope with poor harvests and inadequate range resources. There was heavy reliance on food aid assistance as well as a significant departure from prevailing cropping and animal husbandry practices. The region's cities, areas of high labor demand, and countries hosting relief camps bore the brunt of the refugee burden, and the food security situation in these areas declined proportionately. In Ethiopia, in 1984, some 6.7 million people, 47 percent of whom were children under fourteen, were suffering from starvation and malnutrition (United Nations 1984c). By 1985, there were 3–5 million people in relief camps primarily located in Sudan, Ethiopia itself, and Somalia, with hundreds of people dying daily (United Nations 1987). Mauritania lost up to 80 percent of its livestock (United Nations 1984c), a major blow to herds already dramatically reduced by drought a decade earlier (Franke and Chasin 1980). Meanwhile, regional grain producers such as South Africa and Zimbabwe were forced to halt exports to their needy neighbors and seek out sources of food aid to alleviate their own grain shortages (Robbins 1983).

In 1982 and 1983, drought conditions were so generalized (see Nicholson 1993, fig. 6) that only three African countries had registered significant increases in their food output: Botswana, Lesotho, and Swaziland (United Nations 1984a). In 1984, twenty-seven countries in Africa were experiencing abnormal food shortages, and an additional nine countries were experiencing drought (United Nations 1984b). In 1985, the situation worsened in many areas. Twenty countries (Angola, Botswana, Burkina Faso, Burundi, Cape Verde, Chad, Ethiopia, Kenya, Lesotho, Mali, Mauritania, Mozambique, Niger, Rwanda, Senegal, Somalia, Sudan, Tanzania, Zambia, and Zimbabwe) remained on the UN's critical list, which designated areas where conditions had "sharply deteriorated." According to one report, "some 30 million of 150 million persons living in drought-affected nations [were] categorized as 'seriously affected,' and 10 million of these . . . had to abandon their homes and lands in search of food, water and pasture" (United Nations 1985a).

September of 1985 brought the first faint glimmer of hope. As the *UN Chronicle* announced:

> Eleven African countries—Angola, Botswana, Burkina Faso, Cape Verde, Chad, Ethiopia, Mali, Mauritania, Mozambique, Niger and Sudan—received near or above-average rainfall in the last two months, and people have begun planting. Six other countries—Burundi, Lesotho, Rwanda, Senegal, Somalia and Tanzania—will need to import food commercially, but they might not need emergency food assistance. . . . Also, Zimbabwe had a bumper harvest and Kenya and Zambia are not expected to seek emergency food aid. (United Nations 1985b)

These first rains were interpreted by much of the international community as an indicator that the worst had passed, but, as the UN Office of Emergency Operations in Africa report noted, "One good rainy season can hardly be expected to undo the damage of several years of drought" (United Nations 1985b). Indeed, over $1 billion was still required in 1986 to meet "emergency" needs (United Nations 1987). Six countries remained on the list of those most "critically affected"—Ethiopia, Sudan, Angola, Mozambique, Cape Verde, and Botswana (United Nations 1986).[12]

When drought and famine conditions struck the Sahel region in the early 1970s, the official development community and the international aid organizations were slow to act, and the consequences were especially devastating (Franke and Chasin 1980). When these conditions recurred in the Sahel, the Kalahari, and especially in the Horn, in the 1980s, an all-out effort was made to mobilize support from the international community. What emerged in 1984 and 1985 was considered the "greatest peace-time relief operation in history" (United Nations 1986). Fifty international relief agencies were operating in Ethiopia alone (Vestal 1985). Over the eighteen months between October 1984 and April 1986, official aid to drought and famine affected areas totaled an estimated $3.38 billion, with 6 million metric tons of food aid delivered (United Nations 1987) and 3 million lives saved (United Nations 1986).

The funds generated through aid channels were used initially for such precious commodities as food, shelter, fresh water, sanitation facilities, clothing, blankets, and medicines, but the effort soon expanded to include the revitalization of much of the transportation infrastructure in hard-hit areas and the provision of equipment—train cars, trucks, air transport services—necessary to move the relief supplies to the areas where they were most needed.[13] In effect, the scope of the problem gave rise to a new geography of relief operations. In order to stem the out-migration from rural areas and reduce the oppressive conditions in the major refugee camps, several major relief agencies resolved that food should be "brought to people where they are, before they become destitute, in order that they may remain on their lands and withstand the crisis in familiar surroundings and with the support of their families without the trauma of moving from their homes to new environments" (United Nations 1985a). This entailed moving into areas that were often quite remote, especially in Ethiopia, where some of the worst-affected areas were in the highlands (Vestal 1985). While infrastructural improvements were key, overall the approach was considered "more humane and more cost-effective than resorting to supporting people in camps or in overcrowded urban settlements" (United Nations 1985a).

Faced with these requirements, nearly all of the major donors sharply increased their bilateral aid allocations to Africa. Official aid from the United States to Africa increased by 175 percent in 1984, with a doubling of aid to Ethiopia alone. Italy, Belgium, and West Germany allocated 90 percent, 75 percent, and 40 percent of their total development assistance budgets, respectively, for relief measures in Africa in 1984. At the same time, special dispensations of grain were sent by France and Canada, and additional large donations of cash and in-kind contributions originated in the Netherlands, the European Community, the United Kingdom, Australia, the Soviet Union, and East Germany (United Nations 1984c).

These massive military-style mobilizations of human and material resources through official channels were matched by the broad-based financial support of private citizens. Several spectacular benefits featuring film and music celebrities helped raise public consciousness of the Ethiopian crisis. Some estimates put the total raised by organizers of the internationally telecast rock concert, "Live Aid," at $50–70 million; the special appeal to United States citizens, "USA for Africa," raised $50 million; and a second effort organized by musicians, "Band Aid," earned $12–25 million (Vestal 1985). Much of this aid was subsequently funneled into the programs of private voluntary and non-governmental organizations serving on the front lines of the relief effort. These agencies generated additional public support by making direct appeals to the public via television, radio, and direct mail solicitations. The success of these appeals was illustrated by the case of the United Kingdom, which provided $46 million worth of official funds for drought aid and refugee problems in Africa in 1984. By comparison, in November of the same year, private citizens in the UK generated $10 million worth of donations in a two-week period alone (United Nations 1984c).

SAVING THE CHILDREN

The implications of this outpouring of financial support stretched well beyond the domain of famine relief. While the primary focus of media attention was Ethiopia, the funding streams flowing into Africa had a much broader geographical impact. Development agencies across the continent were suddenly relatively flush with money earmarked for programs that would improve food security, reduce environmental degradation, or contribute toward greater economic stability. In the broadest terms, this meant a shift in emphasis from relief to rehabilitation. Africa as a whole had been dramatically exposed to the world as the site of greatest need. It was, in the words of the UN Secretary-General, "the only continent where standards of living have declined in the past decade and continue to decline today"(United Nations 1986). Confronted with this record of developmental failure, the "development apparatus" (Ferguson 1990) went into overdrive in an attempt to "provid[e] the foundation for resumption of economic progress" (United Nations 1986). More specifically, this post-drought developmental mandate created significant openings that the WID and, later, the gender and environment lobbies were able to exploit to good effect.

At an ideological level, the maternal constructions motivating women-centered development initiatives dovetailed neatly with the multidimensional effort to intervene to save "starving African children" that emerged in the wake of the drought. In practical terms, post-drought reconstruction efforts in particular emphasized the virtues of women's income generation and food production programs as the means most likely to serve the needs of the entire household, and by extension the development needs of the entire continent. By similar token, women were recognized for their contributions to environmental management and were accordingly incorporated into programs designed to renew the natural resource base. In this way, the continent was simultaneously stripped of geography—all of Africa was Ethiopia—and gendered in a very particular way—women became the continent's salvation.[14]

In sum, the 1980s saw the introduction of very specific ideas of maternity under the guise of WID and gender- and environment-oriented development initiatives. These constructed maternities reinforced each other and meshed with the dramatic changes in the region's economic, political, and social geography produced by persistent drought circumstances. The combined effect was a sharp upsurge in international aid which targeted with increasing specificity the Africa region, agricultural and natural resource development, and female producers. Thus, the unprecedented social and economic changes wrought by the Gambia's garden boom took place on an ideological and political economic landscape that was itself undergoing rapid transformations. Rural Gambians were accordingly presented with a particular mix of opportunities and constraints embedded in shifting development policies. What I present in the following chapters is an account that explores in greater detail the structural conditions developers set in place and analyzes the tactics and strategies different groups of rural Gambians developed to manipulate these structures for personal gain. In chapter 2, after a brief introduction to my Gambian research setting, I begin this task with a discussion of the impressions North Bank residents themselves had of their changing economic circumstances.

CHAPTER 2

The Rise of a Female Cash Crop

*A Market Garden Boom
for Mandinka Women*

GENDER AND AGRICULTURE
IN THE GAMBIAN RIVER BASIN

The phenomenon of a cash-crop system managed exclusively by women had little precedent in The Gambia prior to the market garden boom. For that matter, the ethnographic record shows relatively little cash-crop activity by women elsewhere in the West Africa region. Thus, the emergence of a viable market gardening sector in rural Gambia begs several questions. Why did cash-crop production by Gambian women flourish when similar efforts by women in other parts of the region routinely met with failure? How can we account for the timing of the boom, and the fact that it flourished on the North Bank in particular? To what extent was the boom donor-driven, initiated by North Bank residents, or the product of other social, political, or economic factors? How did North Bank residents themselves view the boom? Did women and men see the boom's origins in the same light? Providing answers to these questions will require some background on the history of agrarian change in The Gambia and an explanation of how the basic structural patterns characterizing gender relations in Mandinka society first took shape.

A small strip of land 20–50 km wide and 325 km long, The Gambia is situated between 13 and 14 degrees north latitude on the West African coast. Its population of roughly 1.25 million (1997 estimate) is comprised of three main ethnic groups, the Mandinka (40%), who are the main focus of this study, the Fullah (19%), and the Wolof (15%). Islam is the dominant religion practiced in all parts of the country, although a sizable Christian minority exists in the capital city of Banjul and in Jola-speaking areas on the South Bank. After successive and at times over-lapping colonization attempts by the Portuguese, Dutch, French, and British from the seventeenth to the nineteenth centuries, the land mass of 10,690 square kilometers that makes up the republic today was firmly established as a British colony and protectorate in 1889. The nation achieved formal independence from Britain in 1965.

The Gambia's climate is bi-modal with an eight-month dry season extending from November to June and a four-month rainy season from late June to October. Long-term annual rainfall averages range from 800 mm in Wuli District in the northeast to over 1,100 mm in the southwest in the vicinity of the capital city of Banjul. This record is extremely variable with respect to both geographical and historical trend lines, however, and some analysts argue convincingly for the use of lower estimates reflecting conditions dating back to roughly 1960. Mann (1990) uses a thirty-year record for annual rainfall estimates ranging from 500 mm in the eastern end of the country to 800 mm in the west. Schindele and Bojang (1995), using official Water Resources Department statistics, calculate a range from 750 mm in the east to 950 mm in the west.

The River Gambia itself is tidal for most of the length of the country and saline for a stretch of nearly 100 miles inland. This biogeographical feature is quite significant from a food security standpoint. Nearly half the river's length is unsuitable for irrigation purposes. Thus, unlike central Gambia, where Dey (1981, 1982) and Carney (1988a, 1988b) carried out their studies of large-scale irrigation schemes devoted to rice cultivation, production of rice in the lowland areas of the western half of the country is strictly rainfed. Moreover, under such conditions, substantial rainfall totals are necessary to counteract the effects of salt intrusion on rice plots. Due to prolonged drought conditions, thousands of hectares in this area, which includes Kerewan, have gone out of production, and food security in western Gambia has been undermined as a consequence.

The primary allocation of male and female labor resources in the Mandinka villages that comprise the North Bank garden district has, until recently, been constructed along lines designated by production: (1) of specific crops; (2) within particular spatial domains; (3) during given seasons; and (4) for returns of differing value (table 1). A simplified profile of the gender division of labor would indicate, accordingly,

TABLE 1   DIVISION OF LABOR BEFORE
THE GARDEN BOOM, KEREWAN, NORTH BANK,
CA. 1970

Season	Food Crops Grown by Men	Cash Crops Grown by Men	Food Crops Grown by Women	Cash Crops Grown by Women
Rainy	coarse grains	groundnuts	rice	—
Dry	—	—	vegetables	—

SOURCE: R. Schroeder, field notes.

that men grow groundnuts and the coarse grains (millet, sorghum, and maize) on upland fields during the rainy season and that their domination of groundnut production, the country's main source of foreign exchange, translates into control over most of the cash income generated through agriculture. The profile would also emphasize that women grow rice and vegetables in swamps and low-lying areas, that their gardens constitute the only significant dry-season activity in the traditional Mandinka labor calendar, and that the bulk of the produce they grow is strictly for home consumption (for a more nuanced discussion of this topic, see Boughton and Novogratz 1989; Carney and Watts 1991).

The historical roots of this division of labor date back to the introduction of commercial groundnut production in the 1830s. Prior to that time, a gender division of labor based on individual agricultural tasks rather than crops per se distributed the responsibilities for food provisioning and the opportunities for commodity production and exchange somewhat more evenly among men and women than is the case today (Carney and Watts 1991; Weil 1986). Groundnuts were grown almost exclusively for home consumption in backyard gardens. Locally produced cotton cloth was a principal medium of exchange, and rice grown by women was the featured crop in colonial administrators' plans to send exports out of the river basin (Carney and Watts 1991). With the promotion of market-oriented groundnut cultivation by trading companies, a sharper division of the cropping system into gendered production domains took place. This was partly due to the use of seasonally migratory laborers ("strange farmers") on the groundnut crop and the effect these migrants had on the size and scope of the household unit. Migrant workers provided labor under sharecropping arrangements which stipulated that they be fed by their patron-hosts (Gamble 1949; Swindell 1980). Thus, not only did the "strange farmers" constitute a burden for women in terms of cooking chores, but they also increased the demand for food within the household. With the expansion of groundnut production, production of coarse grains by Mandinka men declined, and rice grown by women became the primary Mandinka staple. Any surpluses women rice growers might otherwise have generated for export were diverted instead toward reproducing the groundnut labor force (Carney and Watts 1991; Watts 1992). During this period, food deficits appeared at the aggregate level (Carney 1986; Watts 1993). The need to generate cash for purchases of supplemental grain increased proportionately, and men plunged ever deeper into groundnut production,[1] even as women intensified rice growing in lowland swamp ecologies. Thus, agricultural production in general became increasingly polarized, both spatially and in terms of gendered labor organization.

While some regional variation exists, this pattern, which conforms closely to the classic male cash crop / female food crop dichotomy exhibited in many parts of sub-Saharan Africa (Guyer 1980; Hemmings-Gapihan 1982; MacCormack 1982; Muntemba 1982; Schoepf and Schoepf 1988) is still in evidence today (Department of Planning 1991). Any impression that the division of labor into male and female production domains in The Gambia is rigid and unchanging would be misleading, however. Several studies have highlighted the fact that unfavorable market and climate conditions since the mid-1970s have prompted male farmers to partially withdraw from cash-crop groundnut production in favor of returning to the cultivation of upland grains (Department of Planning 1991; Jabara 1990; Kinteh 1990; Posner and Gilbert 1987). Conversely, Weil (1986) notes a move by women farmers in Wuli District of Upper River Division in eastern Gambia *into* groundnut production in an attempt to generate cash incomes for grain purchases. Two particularly dramatic changes have grown out of this conjuncture: the first involved incursions by men into the nominally female realm of rice production in response to the introduction of pump irrigation in MacCarthy Island Division (MID) (map 1; Carney 1986, 1988a, 1988b; Carney and Watts 1990, 1991; Dey 1981, 1982, n.d.; von Braun and Webb 1987; von Braun et al. 1990; Webb 1984), and the second was the garden boom.

RICE PRODUCTION AND HOUSEHOLD POLITICS

Traditionally, rice production in MID in central Gambia was organized on the basis of a pattern of reciprocal obligations between male and fe-

[handwritten notes at top of page: "- western ideal that more is always better / - Produce more = happiness"]

male, and junior and senior, members of family units. Household heads often "compensated" junior household members for their labor on "household" rice plots *(maruolu)*—namely, plots managed by household heads for the purpose of meeting joint consumption needs—by providing access to personal land allotments *(kamanyangolu)*. The junior family member was then free to use this allotment for food or cash cropping at his or her discretion.[2] At times, this arrangement gave rise to complex intra-household negotiations. Dey (1981) indicates, for example, that women who grew rice on private *kamanyango* plots in the late 1970s actually sold their rice to their husbands, who then returned the rice to the women to cook for their families.

The introduction of pump-irrigated rice cultivation via projects funded first by the colonial government in the late 1940s, and then by a succession of international donors since the 1960s, seriously disrupted the reciprocal relationship between senior and junior household members. Each of the large irrigation schemes involved land surveys, technical interventions such as plot leveling, and, ultimately, the reallocation of land use rights. Women rice growers frequently lost control of their *kamanyango* plots altogether in this process, as the newly surveyed perimeters were declared *maruo* land under the control of male household heads (Carney 1986; Carney and Watts 1990). In effect, the project-driven change in tenure relations expanded male household heads' claims to their wives' labor and gave them effective control over the surplus product generated on the newly upgraded rice plots. *[handwritten margin notes: "family relations disrupted" and "men get more control"]*

It was not until the mid-1980s, with the growing influence of WID ideologies, that donors explicitly acknowledged the land use rights of female rice growers. In a belated effort to redress some of the more egregious policy failures of the previous decades, donors sponsoring a large-scale irrigation project built in the vicinity of the villages of Jahally and Pacharr insisted in the face of opposition from Gambian managers that plots be registered in women's names (Carney 1986; 1988b). This was a superficial arrangement, however. Village leaders successfully argued that, regardless of who held title to the rice fields, plot rights had to be vested in the household; any other arrangement would run the risk of lands being alienated from the village lineages that originally cleared and claimed them in the case of divorce (see Carney 1986, 1988a). Thus while women were granted *nominal* landholding rights, household heads remained in a position to assert control over the irrigated plots just as they did for *maruo* lands more generally.

In response to the repeated breach of reciprocal norms and dispossession of their land use rights, many women rice growers in the project area simply withheld labor from their respective households and opted instead to enter a growing wage labor market for female rice workers (Carney 1988a; Carney and Watts 1990). In effect, these women replaced their incomes from rice sales with wages earned on gang labor, sometimes by working directly for their own husbands. Many household units were unable to sustain these labor costs and were thus left without the labor resources necessary to meet the demands of dry season double-cropping. This, in turn, called into question both the technical foundation and the economic feasibility of the entire production system, which has since failed to produce at but a fraction of its capacity (Carney and Watts 1990). What the Jahally-Pacharr project demonstrated, then, was that international development interventions geared toward gender equity objectives, but insufficiently attuned to the micro-level power dynamics of the household, were destined to fail.

GAMBIAN HORTICULTURAL SYSTEMS

Somewhat obscured against the backdrop of the international attention received by the Jahally-Pacharr Project was the market garden boom, a gendered development intervention that had a much broader, and arguably more significant, geographical impact for women across The Gambia. There are four main forms of horticultural production in The Gambia: (1) "kitchen" gardens; (2) petty commodity production in rural communal gardens; (3) contract farming; and (4) wholly commercialized production by large-scale capitalist growers (Carney 1992; Daniels 1988; DeCosse and Camara 1990; Jack 1990; Yarbo and Planas 1990). In terms of the gender division of labor, all four forms rely heavily on female labor. Only the first two, however, leave production decisions and the disposition of surplus strictly in women's hands. Briefly, these forms can be characterized as follows.

The category of *kitchen gardens* embraces subsistence gardens located in and around residential compounds and on plots in rice swamps that dry out on a seasonal basis. These gardens are characterized by individualized cultivation, temporary fences and wells that are privately held and maintained (where they exist at all), diverse crop mixes produced from local seed stocks and planting materials, and little or no production for trade. Gardens in residential locations are often cultivated during the rainy season when livestock are tethered to protect field crops and growers are able to rely directly on rainfall for irrigation purposes.

Contract farming of vegetables is confined to the Western Division, where capitalist growers and commodity traders have links to European markets that provide an outlet for select vegetable varieties. Perimeters are equipped with boreholes and hand pumps provided by multilateral donors such as the UNDP, the European Community, and the Islamic Development Bank. Horticultural projects are organized along lines similar to those of other communal gardens, with the exception that production decisions are firmly dictated by contract and inputs are provided directly to female growers on credit. Costs are recovered at the end of the season before the contracted growers are paid, and returns to labor are typically quite low.

Capitalist horticulture is found in fifteen or sixteen large-scale farms (5–150 ha), all controlled by men and located in the Western Division. Crops grown include a number of "exotic" vegetables that are either exported directly to Europe or destined for sale locally to tourist hotels. Growers hire female wage labor for select tasks and utilize the most advanced drip, sprinkler, and ditch irrigation techniques.

My references to the garden boom in this study are exclusively directed at the fourth form of horticultural production, vegetable gardens maintained by *rural petty commodity producers*. In particular, this study emphasizes Mandinka vegetable growers along The Gambia's North Bank. In recent years, development projects organized by NGOs and voluntary agencies have supported the intensification of women's vegetable production in some 300 communities throughout the country. Unlike kitchen gardens, these enterprises are sited on the outskirts of villages or along the fringes of rice swamps. They are typically communal in organization, insofar as usufruct rights to land are granted to groups of women as opposed to individuals. Fence maintenance and crop protection are also communal responsibilities, although production per se is typically organized on an individual or small work group basis. Garden projects often originate independent of outside support but almost inevitably attract donor grants for fence and well construction once they are established. Crops are often grown from imported hybrid seed, and planting mixes are determined largely by market demand. Unlike kitchen gardens, the bulk of the crop is grown for sale, and unlike contract farming and capitalist production, the proceeds from these sales are controlled by the women who cultivate the crops. Since the labor required to maintain vegetables is intense, most such production is carried out during the dry season to avoid competition from the rainy season rice crop.

Conventional wisdom has held that returns to women's vegetable production in rural areas (i.e., parts of the country outside the Western Division) are meager, and the marketing problems women gardeners face insurmountable, given regular gluts in rural markets during the peak sales season (Vakis 1986). While it is true that gardeners in most parts of the country have great difficulty marketing their produce, the lack of attention to vegetables grown outside the immediate peri-urban area surrounding Banjul owes as much to the urban bias of the international and state-funded research apparatus as it does to any inherent problems in the vegetable trade itself. As a consequence of this bias, the North Bank and other production districts outside the immediate urban area received scant attention in governmental action plans drawn up in the 1980s in support of women's garden production. The horticultural component of the $15.1 million World Bank Women in Development program, for example, was concentrated primarily in the Western Division.[3] Part of the original motivation for my own study was accordingly to bring the remarkable accomplishments of the North Bank gardeners to light, and thereby prompt the Gambian government to give rural women horticulturists their due (Schroeder 1991a; 1991b; 1991c).

ORIGINS OF THE GARDEN BOOM

Like the Jahally-Pacharr Project, the impact of intensified vegetable commodity production on rural areas such as the North Bank has been complex and multifaceted. To take a banal example, when a Kerewan woman was asked directly as part of a structured survey of market gardeners to explain the origins of the garden boom, she responded that the key was growers' access to cabbage (fig. 2) and capsicum pepper seed. This reply highlighted two of the leading horticultural crops grown in Kerewan, neither of which was extensively produced in the area prior to the garden boom. It implied a shift away from a horticultural regime based almost solely on locally produced seed and planting materials (e.g., shallot bulbs), and a transition toward one based on vegetable hybrids produced abroad. It thus underscored deepening market penetration and the prominent role of international development agencies in rural Gambia, concomitant increases in market dependency for input supplies, and the displacement of local technology. Furthermore, the introduction of cabbage and capsicum peppers afforded growers significant new options in crop management. The respective growth patterns (i.e., plant height, leafiness, spread, duration, etc.) of the two new crops

Figure 2.   Diversification of Mandinka horticulture. The introduction
of cabbage production into the Mandinka horticultural system meant
heavier dependence on foreign inputs but gave gardeners greater flexibility
in producing a more diverse and economically viable selection of crops.

helped growers maximize productive output in both spatial and seasonal terms while sustaining a steady cash income (see chap. 3). Finally,
the prominence of cabbage and capsicum production implied changes in
demand for the new foodstuffs that became available with the boom. As
one man put it: "Women used to prepare meals without many vegetables. We were not spending physical cash for daily meals. It is [only]
now that someone would refuse a meal without vegetables."

These somewhat differing views of cabbage production belie subtle
differences in the ways men and women interpreted the garden boom.
Generally, gardeners explained the boom phenomenon on the basis of
changes in production factors, whereas their husbands seemed more
inclined to stress consumption. Thus, for many women, climate change
and assistance from outside funding sources were key explanatory vari-
ables. By contrast men often insisted that the roots to the garden boom
lay deeper, in the generalized commercialization of the rural economy
and the North Bank's incorporation into national and international
market networks. In their view, new tastes and tolerance levels necessitated higher cash incomes. It was the imperative to meet these socially
defined needs that resulted in the expansion of the horticultural sector.
Neither of these positions is necessarily consistent with conventional expectations regarding the maternal altruism of women, or the essentialist
assumption that men are wholly preoccupied with production. As such
they are worth exploring further.

PRODUCTION DYNAMICS: CLIMATE CHANGE
(1970–1990), GENDER AND DEVELOPMENT

There is no rice; the rains don't come. There are no groundnuts; the
rains don't come. Everything comes from the garden now. But the
women are tired.   *Kerewan gardener*

Several Kerewan gardeners interviewed in connection with this study
emphasized that the garden boom had occurred by default. The production levels of other crops, and thus the resiliency of the rural agrarian economy in general, had declined sharply under recurrent drought
conditions, and women were forced to intensify vegetable production
in response to these changes. When drought conditions swept the continent in the early 1980s, The Gambia River Basin was not among the
worst affected areas. Unlike its West African neighbors, Mauritania,
Senegal, Mali, or Burkina Faso, the country was not on any of the UN's

"emergency need" lists. There was minimal displacement of the rural population and little blatant starvation. Indeed, when funds raised for emergency relief by celebrity groups were distributed, The Gambia did not even qualify for support.

These facts notwithstanding, food production conditions in The Gambia declined significantly between 1970 and 1990. Norton et al. (1989) reported in 1989 that average annual rainfall had declined 24–36 percent in different parts of the country over the previous two decades; over that same period the rainy season itself had decreased in length by 14–24 days (see also Mann 1990). Salinization of rice swamps was widespread, as noted above, and the quality of forest cover on the uplands declined significantly in many areas (chap. 6). The garden boom accordingly took place against the backdrop of an unambiguous downward trend in rainfall averages and the degradation of several key resources.

In terms of specific effects on crop production, Posner and Gilbert (1987) note that most of the rain shortfall in The Gambia occurred in the month of August, which is normally the heaviest rainfall month: "Poor rainfall at this time prevents recharging of the soil profile for subsequent use in October[;] reduces run-off, markedly affecting the possibly of growing swamp rice[;] and can result in short dry spells for the early cereals (*sunno* millet, maize)" (Posner and Gilbert 1987, 3). A comparison of the colonial research record with contemporary practice sheds further light on the situation. Gamble's (1955) account of agricultural activity in Kerewan in the late 1940s includes a detailed calendar of events for the period from May 1948 to April 1949.[4] He indicates that the first heavy rain storms of 1948 came during the first week of June, with women beginning hoeing operations in their rice fields immediately thereafter. By contrast, in the late 1980s, the unreliability of the onset and subsequent continuation of seasonal rains, as well as the absolute decline in early season rainfall totals, effectively prevented ground preparation and planting operations in rice fields much before the middle of July.

Gamble's account notes further that men in Kerewan grew very little maize and no early-maturing millet *(sunno)* in 1948. September found women continuing to transplant their long-duration rice crop in deep water plots, some of which would not be harvested until late December, at which point women's involvement in vegetable production proceeded in earnest. Underscoring the differences between the two periods, Posner and Gilbert (1987) note that early-maturing millet, a drought-adapted crop, had almost entirely replaced sorghum and late millet in most crop rotations and that the area devoted to millet production had increased relative to groundnuts. The switch to early millet illustrates the fact that a number of shorter-duration, drought-resistant, or cash-earning crops were incorporated into the Mandinka cropping calendar in the 1970s and 1980s, a trend that included vegetables grown in market gardens. Posner and Gilbert (1987) also indicate that the practice of transplanting rice, normally associated with long-duration varieties grown under flooded or extremely moist soil conditions, was all but discontinued on the North Bank as recurrent droughts altered production conditions in lowland areas.

The reconfigured cropping pattern altered seasonality of labor demand, and this had direct implications for vegetable production. The adoption of shorter-duration rice varieties meant that rice could be harvested as much as two months earlier than long-duration varieties women grew under earlier conditions.[5] This change effectively freed women from some of the burdens rice production placed on their labor and allowed them to devote a significantly greater proportion of their labor to their private enterprises instead of producing solely for joint consumption. In addition, by planting vegetables in September / October as opposed to November / December, gardeners were able to take advantage of more favorable growing conditions—cooler temperatures, higher ground water tables, more residual soil moisture—and better harvest prices (Daniels 1988).

While secular climate changes related to drought produced certain opportunities for vegetable growers, they also placed a strain on household finances. Rural households were forced to diversify sources of income and food stocks in a variety of ways in order to reproduce themselves under difficult circumstances. The drive to intensify horticultural production must also be seen in this light, as women pushed to expand their gardens out of necessity. Consequently, when agencies armed with capital and a definite mission to redress gender inequities sought to establish themselves in the country in the late 1970s and 1980s (table 2), the increasingly viable horticultural enterprises attracted their attention.

WID-oriented developers found rural Gambian women to be both industrious and overburdened. Some estimates indicated that women farmers in The Gambia were providing as much as 75 percent of domestic food production: they grew some 90 percent of the country's rice, the principal staple crop, owned 40 percent of the country's sheep and goats, and controlled most of the country's poultry flock (Norton-Staal

TABLE 2   FOUNDING DATES OF MAJOR
INTERNATIONAL NGOS, MISSION GROUPS, AND
VOLUNTARY ORGANIZATIONS ACTIVE IN
HORTICULTURAL PROGRAMMING IN THE GAMBIA

Founded	NGO / Mission Group / Voluntary Organization
1964	Catholic Relief Services
1973	Canadian University Services Overseas
1977	CARITAS
1978[a]	Methodist Mission Agricultural Program
1978	Voluntary Service Overseas (UK)
1979	Action Aid, The Gambia
1982	Save the Children Federation/USA
1982	U.S. Peace Corps Small Projects Assistance
1984	Christian Children's Fund
1984	Freedom from Hunger Campaign
1985	The Good Seed Mission

SOURCE: Campbell and Daniels 1987; Mann 1990; TANGO n.d.
[a]This date markes the initiation of MMAP's intensive well-digging program and
the establishment of a tree nursery in Brikama that was to figure prominently in agroforestry promotional efforts in the mid-1980s (see chap. 6).

1991). On the negative side, developers also determined that Gambian
women had acute health needs. For example, in 1990, the maternal
mortality rate in The Gambia was 10 per 1,000 live births, with some
areas reporting rates as high as 20–22; infant and child mortality stood
at 143 and 242 per 1,000, respectively, and the overall fertility rate was
6.5 live births per woman.[6] These rates were all among the highest in
Africa (Cleaver and Schreiber 1993). In short, when most nongovernmental agencies initially entered the Gambia in the late 1970s and early
1980s, they found a gendered economy that corresponded directly with
the Boserup (1970) profile that had helped inspire the original WID con-
ference in Mexico City in 1975.

NGOs, mission groups, and voluntary organizations saw the introduction of garden projects as the ideal intervention for addressing the
multiple needs inherent in the structure of The Gambia's agrarian economy. In theory, produce grown within the gardens served a dual purpose
of meeting dietary needs while generating cash incomes.[7] Garden projects were also often favored by the agencies involved because they
generated immediate, visible, and productive changes in the landscape.
Consequently, as drought rehabilitation aid and WID funds funneled
through these agencies into The Gambia, gardens became a high priority, and hundreds of grants for barbed wire, tools, hybrid seed, and well-

TABLE 3   DIVISION OF LABOR AFTER
THE GARDEN BOOM, KEREWAN, NORTH BANK,
CA. 1985

Season	Food Crops Grown by Men	Food Crops Grown by Women	Cash Crops Grown by Men	Cash Crops Grown by Women
Rainy	coarse grains	rice	groundnuts	—
Dry	—	—	—	vegetables

SOURCE: R. Schroeder, field notes.

digging costs were quickly negotiated. WID programming by international NGOs and voluntary organizations thus provided an ideological and material basis that underpinned the nascent garden boom through much of the ensuing decade.

It is important to note that the garden boom marked the emergence of a new regime in gender / environment relations along the North Bank. For the first time, the Mandinka cropping calendar included a period of intense agricultural activity during the dry season (table 3). Horticultural efforts were directed at reclaiming low-lying land resources that had been greatly marginalized because of the geophysical processes accompanying recurrent droughts (see the parallel discussion in Carney 1993). The success of these efforts was premised, however, on a shift from rainfed agriculture to irrigated production. In a cultural context where adult men rarely draw water from wells for any reason (see discussion in chap. 3), the success of the transition hinged directly on the use of female labor. The net effect of these changes was that the incomes generated through dry season market gardening became critical to household reproduction as drought conditions and national economic policies forced male farmers to shift part of their energies from producing groundnuts for cash to producing millet for home consumption.

CHANGING CONSUMPTION PATTERNS
AND MARKET INTEGRATION

I have stressed that women gardeners seemed predisposed to analyze the garden boom in terms of production dynamics. This is not to say that they were completely oblivious to the connections between increasing commodification and their own efforts to generate cash incomes. To the contrary, they were often eloquent in expressing their concern over what they saw to be an increasingly desperate household budgetary situation. Nonetheless, it was their partners, the male residents of the North Bank garden district, who seemed most impressed by the changes in the actual patterns of consumption accompanying the garden boom: new dietary preferences, lower tolerance for substandard housing and water supply sources, higher transportation costs, more lavish ceremonial practices, and larger wardrobes of clothing and jewelry. Many men insisted that the new tastes were causes, and not merely outcomes, of the boom. They stressed that newly created needs had increased pressure on family finances and driven individual household members—women in particular—to intensify income-generation efforts in order to satisfy personal requirements.

materialism associated w/ development

The reasons behind the evident predisposition of many of my male informants toward these matters are threefold. First, food costs accounted for well over half of all the cash spent in rural areas (Jabara et al. 1991), and it was the men who were customarily responsible for procuring supplemental grain in the event that domestic supplies failed to last from one harvest to the next. Production figures for rice, the staple of choice for most ethnic groups in The Gambia, fell significantly due to declining rainfall, including an apparent drop of some 50 percent over the period 1972–1987 (table 4; see Kinteh 1990). Partially in response to this shortfall, rice imports soared (table 4), and consumer prices increased sharply (Jabara 1990).[8] Official prices for imported rice doubled between 1980 / 81 and 1984 / 85, and more than doubled again in the four years following the implementation of structural adjustment reforms in 1985 / 86 (Jabara 1990).[9] Since men were at least nominally responsible for grain purchases under these conditions, it stands to reason that they would interpret the boom primarily in terms of the need for additional cash at the household level.

1)

The second factor of particular relevance to aspects of household affairs typically managed by men was the deepening integration of the North Bank into international markets. A fairly intense period of feeder road construction and trading center establishment along the Gambia-Senegal border coincided exactly with the early stages of the garden boom. Improvements on the Upper Niumi District road from Buniadu to Kuntair (1981–1984), and new feeder roads to Kerr Musa Saine, Jowara, Salikenye, and Katchang (1985–1987), helped link important vegetable-growing districts to weekly markets *(sing. lumoo;* pl. *lumoolu)* in Ndungu Kebbe, Kerr Musa Saine, Kerr Pateh, and Farafenye, all founded around 1982 (see map 2).[10] The facilitation of cross-border

2)

TABLE 4   CEREAL PRODUCTION AND IMPORTS, THE GAMBIA: 1972–1987

(,000 metric tons)

Commodity	Year						
	1972–1974	1980–1982	1983	1984	1985	1986	1987
Milled rice	28.7	49.8	56.4	78.1	93.5	103.1	137.4
Domestic production	18.6	18.3	16.6	12.9	13.4	11.4	12.0
Imports	9.9	26.4	29.7	50.0	64.8	73.8	113.0
Food aid	0.2	5.1	10.1	15.2	15.3	17.9	12.4
Wheat	5.3	15.9	10.5	16.2	13.9	23.8	19.6
Imports	4.0	11.1	7.8	12.5	8.6	23.6	19.3
Food aid	1.3	4.8	2.7	3.7	5.3	0.2	0.3
Millet	32.6	17.5	29.5	22.4	32.7	46.4	43.4
Domestic production	32.6	15.0	28.6	21.9	32.7	46.4	43.4
Imports	0.0	2.5	0.9	0.5	0.0	0.0	0.0
Sorghum (domestic production)	—	10.0	13.5	6.0	7.0	10.0	7.6
Maize (domestic production)	1.7	7.2	14.4	7.2	10.6	22.5	14.7
Coarse grains (imports)	3.0	4.8	0.0	0.0	0.6	0.4	0.4
Total cereals	71.3	105.2	124.3	130.0	158.3	206.3	223.1

SOURCE: Adapted from Jabara 1990, 29.

trade had significant implications for rural budgets. On the one hand, the weekly markets served as relatively reliable outlets for vegetable produce moving north to Senegal; on the other, they were an important source of supplies such as seed, pesticide, and fertilizer. More to the point, *lumoolu* provided growers and their families with access to many of the material goods that corresponded with the expanded range of socially defined needs in rural households. Consequently, the costs for naming ceremonies, circumcisions, weddings, and other special occasions, all of which were obligations typically borne by male household heads prior to the boom, increased sharply.[11]

Finally, as if these pressures were not enough to heighten awareness of changing consumption patterns, the capacity of rural male farmers to earn the cash necessary to fill these needs declined sharply just as the garden boom took off. Income from groundnut sales stagnated or declined between 1978 and 1988 (Jabara 1990), the only respite coming in the form of back-to-back increases in producer price in 1985 / 86 and 1986 / 87. These were designed specifically to offset the most egregious effects of the Economic Recovery Program (ERP) (von Braun et al. 1990) but were quickly overtaken by inflation induced by currency devaluation (Puetz and von Braun 1990). Thus, for example, a nominal increase of 295 percent in the groundnut producer price was reduced to a 67 percent rise in real terms and quickly reverted to pre-ERP levels the following year. At the same time, the effects of the ERP also dramatically altered production costs for groundnut growers. In real terms, fertilizer prices between 1984 and 1987 increased 11 percent, groundnut seed nearly doubled in cost, and the price of hired labor and draft animals rose by nearly a third (Johm 1990; Puetz and von Braun 1990).[12] The net effect of these changes was decisive: a sharp decline in fertilizer usage,[13] lower earnings for groundnut producers on the whole (Jabara 1990), and a partial withdrawal of male labor from cash-crop groundnut production in favor of early millet (Jabara 1990; Puetz and von Braun 1990).

In sum, rural male peasant farmers on The Gambia's North Bank were faced with escalating costs for basic goods and services and increasing demands for a wide range of amenities by their families. Moreover, these changing circumstances came at a time when the major source of cash income for male farmers was curtailed due to a classic "cost-price squeeze" (Bernstein 1982). The coincidence of these developments accounted in large measure for men's preoccupation with consumption matters. In practical terms, men turned to staple crop production as one strategy for reducing pressure on the household. They also often simply defaulted on many of their financial obligations to their families. This meant that women were forced to compensate by increasing cash-crop vegetable production, one of the few sectors of the agricultural economy that thrived along the North Bank in the face of the double onslaught of drought and structural adjustment over the course of the 1980s.

CHAPTER 3

Gone to Their Second Husbands

Domestic Politics and the Garden Boom

One of the offshoots of the surge in female incomes and the intense demands on female labor produced by the garden boom was an escalation
of gender politics centered on the reworking of what Whitehead once
called the "conjugal contract" (Whitehead 1981; see also Jackson 1995).
In Kerewan, the political engagement between gardeners and their husbands can be divided into two phases. The first phase, comprising the
early years of the garden boom, was characterized by a sometimes bitter war of words. In the context of these discursive politics, men whose
wives seemed preoccupied with gardening claimed that gardens dominated women's lives to such a degree that the plots themselves had become the women's "second husbands." Returning the charge, their wives
replied, in effect, that they may as well be married to their gardens: the
financial crisis of the early 1980s had so undermined male cash-crop
production and, by extension, husbands' contributions to household
finances, that gardens were often women's only means of financial support during this period.

As the boom intensified, so, too, did intra-household politics. The
focus of conflict in the second phase—which extended into the mid-
1990s—was the role of garden income in meeting household budgetary
obligations. Several studies have examined "non-pooling" households in
Africa, that is, households in which men and women tend to engage in
distinctly different economic activities and control their own incomes
from these enterprises.[1] The garden boom offers a case study in which

women, by virtue of their new incomes, entered into intra-household negotiations over labor allocation and income disposition with certain
economic advantages. The upshot of these negotiations was not, however, quite so simple. In terms of budgetary obligations, women in the
garden districts assumed a broad range of new responsibilities from
their husbands. Moreover, they frequently gave their husbands part of
their earnings in the form of cash gifts. This outcome appears in some
respects as a capitulation on the part of gardeners. I argue in this chapter, however, that it can also be read as symbolic deference designed to
purchase the freedom of movement and social interaction that garden
production and marketing entailed. In effect, gardeners used the strategic deployment of garden incomes to win for themselves significant autonomy and new measures of power and prestige, albeit not always at a
price of their own choosing.

MAPPING MARITAL METAPHORS

It is because wives had nothing to do before except sit near their
husbands. But now wives are running both day and night struggling
for survival. That is why [relations between men and women] ha[ve]
changed. If you want to do something for your husband, you must go
to the garden.   *North Bank gardener*

. . . because what they produce from the rice fields is meant mainly
for home consumption. But what they produce from the garden goes
directly to their personal use, that is why they are more concerned
with gardening. Probably that is why some of them are at odds with
their husbands.   *Gardener's husband*

As noted in chapter 2, the garden boom marked a fundamental shift
away from predominantly rainfed agriculture and toward groundwater-
based irrigated production. Not only were gardeners required to mobilize for a second full production season with the cessation of seasonal
rains, but high evapotranspiration rates during the dry season gardening
period necessitated a rigorous irrigation schedule. Most women watered
their crops twice daily to cope with these adverse conditions. Depend-
ing on the distance between a woman's home and her garden site(s) and
the extent of her individual holdings, this could mean up to six hours a
day spent on gardening tasks alone.

   During the first phase of the boom, many gardeners' husbands on the

North Bank were deeply resentful of their wives for undertaking such a time-consuming activity away from home. With little yet to show for all the effort, the women's absence was simply viewed as a loss of control, and by extension a loss of male prestige. Certain acts were especially charged with symbolic significance. For example, men particularly resented the fact that the women were no longer available to greet guests properly in their homes. As one male informant put it: "Presently you are here talking to me but my wives are not here. They are not doing what is obligatory. If you had found them here, they would have given you water to drink, and perhaps you would need to wash as well. I am now doing . . . what they are supposed to do." The ability to show hospitality to honored guests is an important measure of the host's prestige in Mandinka society. That men were no longer able to call upon their wives and daughters at a moment's notice to perform this service reflected badly on them. At least one North Bank community banned gardening altogether in the mid-1980s because of the shame and irritation men felt under these awkward circumstances (Schroeder and Watts 1991).

In the eyes of the development agents who promoted the garden boom, the work routines followed by rural market gardeners were seen as the embodiment of positive maternal values. In the garden communities themselves, however, they very quickly became imbued with meanings associated with a failure to meet marital obligations. Many men publicly ridiculed the gardens, dismissing them derisively as a waste of time. They also often impugned their wives directly for what they saw as a lack of commitment to their marriages. It became commonplace for men to complain that the women had taken new "husbands." Thus, if a man's wife was busy working in her garden and someone asked her whereabouts, he often replied, "She's gone to her husband's" (Mandinka: "*a taata a ke yaa*"). Indeed, this phrase was so widely repeated that it became a kind of shorthand for marking women's neglect of their marital responsibilities. It demonized gardeners as bad wives.

Just as was the case with explaining the roots of the garden boom, when asked directly to interpret the metaphor that rendered women's gardens into their husbands, men and women in the garden districts offered somewhat distinct readings. One interpretation commonly offered by men reflected their frustration at the fact that gardens dominated women's lives to such a degree that their husbands hardly saw them on a day-to-day basis. According to this interpretation, which was widely acknowledged by gardeners themselves, the gardens supplanted husbands' wishes as the primary ordering force in a woman's workday. As one gardener's husband put it: "A wife is brought home to fulfill her obligations to her husband. She should be around her husband all the time to render such services. In the case of garden work, women are away from home almost the whole day. They do not perform what is required of them." Vegetable growers "greeted" *(saama)* their gardens (and not their husbands) when they watered their vegetables first thing in the morning;[2] they spent their days "at the side of" *(daala)* their gardens; and they brought their gardens water at dusk (i.e., at precisely the time when a man might expect his bath water to be delivered). Consequently, gardeners' marriage partners found themselves increasingly without companionship and forced, by default, to assume new domestic labor responsibilities. This was especially true of older men who had been economically marginalized due to age or ill health and who spent a great deal of time within the spatial confines of the family compound or its immediate vicinity. Indeed, early in the boom, the loss of these "prestige services" caused a great deal of bitter resentment.

In contrast to the impression men left of absentee wives, women in the garden districts often offered an interpretation of the garden-husband metaphor that emphasized the importance of garden earnings in meeting household budgetary obligations. For them, gardens had, for all practical purposes, replaced husbands as the principal source of cash for subsistence and other forms of consumption ("Women are doing what men *should* be doing"). Somewhat sardonically, they maintained that women might just as well be married to their gardens. One grower underscored the point dramatically by asserting that not just her garden, but the *well bucket* she used to irrigate her vegetables was her husband because everything she owned came from it. Playing to the laughter of several women gathered nearby, she spoke with her voice rising in mock rage: "This [indicating her dress]; this [her shoes]; this [her earrings]; this [miming the food she put into her mouth]; and this [clutching her breast to indicate the food she fed her children]—they all come from this bucket! That's why this bucket is my husband!!"

Clearly, the Mandinka marriage system was placed under significant strain due to the changes accompanying the push toward commercialization. It is equally apparent that the hard fought rhetorical struggle in which men and women mapped marital meanings onto garden spaces had as its object the right to occupy the moral high ground in the broader battle over the conjugal contract. A variant phrase used by men marked women's gardens as their "*second* husbands." In this usage, the men's rhetoric invoked the tensions and resentments that accompany a man's taking of a second *wife*. It reflected the fact that, while first marriages are often arranged, second marriages can be undertaken by choice. The usage echoed jealous charges that men give second wives preferential treatment because they consider them prettier and stronger, or because they are more fertile than their older co-wives. In seizing and using the "second husband" metaphor to castigate women, men in garden districts attempted to turn the tables on their wives, to assume a superior moral position from which they could wield leverage in the renegotiation of conjugality that inevitably ensued.

WOMEN'S GARDEN INCOMES: RAISING THE STAKES

The stakes in these negotiations rose sharply once the boom was fully under way. Evidence drawn from several sources indicates that the distribution of the boom's benefits was geographically quite uneven. Carney (1992) reported seasonal incomes in the range of only 200–250 Gambian *dalasis* ($30–35) for rural growers on the South Bank. This figure corresponds with the estimates reported by Barrett and Browne (1991) for women working as contract labor on irrigation schemes sponsored by the Islamic Development Bank in Western Division. By contrast, Nath (1985b) estimated earnings for the five top onion growers on one rural project to have been in the range of $65–130, and my own research in four different North Bank communities in 1989 and 1991 documented income levels that were higher still. On the low end of the range were the villages of Illiasa (D434) and Jumansari Baa (D593) located along the Bao Bolong in Upper Baddibu District. On the upper end were the two most productive communities in the survey, Kerewan (D1096), and Niumi Lameng (D1377).[3] These figures underreported income from tree crops, off-season production, and, in the case of Niumi Lameng, individual sales negotiated outside the village buying point. Bearing these factors in mind, the average gross income for North Bank growers in the four villages studied in 1991 was well over D1000, or roughly $135 at 1991 exchange rates,[4] an amount approaching the annual per capita income estimate of D1500 for all rural Gambians (Jabara et al. 1991).

It is important to consider how net returns differed from the gross earnings figures I have referred to thus far. Unfortunately, detailed figures on the costs of vegetable production are only available for Kerewan, and these may not adequately reflect the economic circumstances of other districts. Table 5 shows that Kerewan growers spent roughly a

TABLE 5   ANNUAL PRODUCTION COSTS
IN KEREWAN GARDENS, 1991 DRY SEASON[a]

Item	Dalasis	Income (%)	Total Costs (%)
Well digging	2,776	2.5	7
Well deepening	2,160	2.0	6
Fertilizers	4,868	4.4	13
Seed	2,325	2.1	6
Pesticide	373	0.3	1
Buckets, rope, tools[b]	3,000	2.7	8
Transport, produce	14,375	13.1	38
Transport, self	8,145	7.4	21
Total costs	38,022	34.7	100
Net earnings	71,623	65.3	
Total income	109,645		

SOURCE: R. Schroeder, survey data.
NOTE: $n = 99$.
[a]These estimates do not include the costs of labor or land; for the latter, see chapter 5.
[b]Imputed values.

third of their income on recurrent expenses in 1991. Over 20 percent of the income earned by Kerewan growers was absorbed by transportation costs. This is a reflection of Kerewan's locational disadvantage relative to markets in Farafenye and / or Banjul, where more favorable prices were consistently offered for produce (map 2; Schroeder 1991a).[5] In addition, Kerewan's situation adjacent to extremely low-lying lands resulted in a proliferation of hand-dug garden wells. The prevalence of relatively sandy soils in village gardens accounted for a high rate of well collapse, and thus the significant recurrent expense associated with well maintenance and replacement. On balance, production expenses in Kerewan appear to have been considerably higher than in the other North Bank villages in the study.

After deducting costs, the earnings of North Bank gardeners seem quite modest, but it would be difficult to overstate their social and political significance. Before the garden boom, men in Mandinka society had powerful economic levers at their disposal which they could, and did, use to "discipline" their wives (Carney and Watts 1991). They controlled what little cash flowed through the rural economy due to their dominant position in groundnut production and were able to fulfill or deny a range of their wives' expressed needs at will. These included such basic requirements as clothing, ceremonial expenses (naming ceremonies, circumcisions, and marriages for each individual woman's children), housing amenities, and furnishings. The power vested in control

over cash income was only enhanced by polygamous marital practices
and the opportunities they afforded to play wives off against one another. A second advantage was derived from the husband's rights in divorce proceedings. In the event of a divorce, Mandinka customary law
requires that the bride's family refund bridewealth payments. Consequently, when marital relations reach an impasse, divorce is not auto-
matic; the financial arrangement between the two families must first be
undone. Typically, the woman flees or is sent back to her family so that
they can ascertain to their own satisfaction whether she has made a
good faith effort to make her marriage work. The onus is on the woman
to prove her case, however, and she is not infrequently admonished by
her own family to improve her behavior before being returned to her
husband.

The advent of a female cash-crop system reduced the significance of
both these sources of leverage, not least because women's incomes had
outstripped their husbands' in many cases. A rough comparison of the
garden incomes of women in Kerewan and Niumi Lameng and the earnings their husbands reported from groundnut sales showed that 81 percent and 47 percent of women in the Niumi Lameng and Kerewan sam-
ples, respectively, earned more cash than their husbands from sales of
these crops.[6] This reversal of fortunes changed fundamentally the way
male residents of the garden districts dealt with their wives:

Before gardening started here, if you saw that your wife had ten dalasis
you would ask her where she got it. At that time, there was no other
source of income for women except their husbands. . . . But nowadays
a woman can save more than two thousand dalasis while the husband
does not even have ten dalasis to his name. So now men cannot
ask their wives where they get their money, because of their garden
produce.   *Gardener's husband*

Indeed, the garden boom reduced male authority ("If she realizes she is
getting more money than her husband, she may not respect him"), and
the extent of gardeners' economic influence expanded proportionately.
The simple fact that women could largely provide for themselves ("If we
join [our husbands] at home and forget [our gardens in] the bush, we
would all suffer. . . . Even if he doesn't give you [what you want], as long
as you are doing your garden work, you can survive") constituted a serious challenge to the material and symbolic bases of male power. In the

first phase of conflict brought on by the boom, men openly expressed their resentment in pointed references to female shirking and selfishness. Their feelings were also made plain in actions taken by a small minority who forbade their wives to garden, or agitated at the village level to have gardening banned altogether (Schroeder and Watts 1991). In the second phase, men dropped their oppositional rhetoric, became more generally cooperative (cf. Stone et al. 1995), and began exploring ways to benefit personally from the garden boom. Sensing the shift in tenor of conjugal relations, women, accordingly, began a prolonged attempt to secure the goodwill necessary to sustain production on a more secure basis.

The key to vegetable growers' success in this regard lay in their strategic deployment of garden incomes. For reasons that I explain below, the disposition of garden income often concentrated in the hands of older women who worked in tandem with their daughters. This was significant, insofar as an older woman's social obligations were likely to be broader than those of a younger woman. In deciding how much of the surplus generated by the work unit would be allocated to each individual member of the group, and what form the compensation was to take, the unit leader shaped the complex politics of the horticultural boom. She chose, for example, whether to buy a bag of rice for her daughter and son-in-law's family, pay for the school expenses of a nephew or grandchild, disburse portions of the cash surplus at season's end to each work unit member, give a cash gift to her own husband, or simply keep the funds to buy personal items for herself (fig. 3). In so doing, this woman accumulated a significant measure of power and prestige, elements that would have accrued exclusively to her husband or other male relative in the past by virtue of his control over the groundnut cash crop.

Rural Gambian households were under significant economic stress in the late 1970s and early 1980s, when the boom was initiated. Given the poor market conditions facing the male cash-crop sector at the time, many men were forced into what might be called *legitimate default* vis-à-vis their customary obligations to feed or otherwise provide for their families. Survey data show that both senior members of garden work units and women working on their own took on many economic responsibilities that were traditionally ascribed to men. Fifty-six percent of the women in the Kerewan sample, for example, claimed to have purchased at least one bag of rice in 1991 for their families.[7] The great majority bought all of their own (95%), and their children's (84%), clothing and most of the furnishings for their own houses. Large numbers

Figure 3. Discretionary income. Gardeners possessed considerable discretion over income derived from vegetable sales; whereas they often displayed "maternal altruism" and gave money to their families, they were equally likely to use their earnings to buy personal items such as radios, furniture, and clothing.

Figure 4. Shifting household budgetary responsibilities. Earnings from
market gardens have outstripped the income men earn from their own cash-
crop sales in many Gambian households. Consequently, women have assumed
a broad range of financial responsibilities from their husbands, including the
cost of feast day clothing.

took over responsibility for ceremonial costs from their husbands, such
as the purchase of feast day clothing (80%; fig. 4), or the provision of
animals for religious sacrifice.[8] Many paid their children's school expenses.[9] In a handful of cases, gardeners undertook major or unusual expenditures such as roofing their family's living quarters, providing loans
to their husbands for purchasing draught animals and farming equipment, or paying the house tax to government officials. There are, once
again, unfortunately no baseline data that could be used to gain historical perspective on this information. Nonetheless, several male informants stated unequivocally that, were it not for garden incomes, many
of the marriages in the village would simply fail on the grounds of "non-
support."

One category of income expenditure by Kerewan's gardeners remains
unexplained. Of the women sampled, 38 percent reported undertaking
some measure of direct support of their husbands via cash gifts; typically dispersed in small, regular amounts, they occasionally amounted
to hundreds of dalasis at the season's end. The question is, why did

women feel compelled to share their income in this way? Were their husbands performing special services such as digging wells, or cutting fence posts, that constituted material support for the garden boom? If so, the payments could be interpreted as a euphemized form of wage payment (Scott 1985). Alternatively, did men help finance their wives' gardens? If so, women's cash gifts might legitimately be construed as a return on the men's investment, a disguised form of interest payment. If, however, neither of these scenarios held true, then how should the cash gifts be interpreted?

A survey of ninety-nine gardeners in 1991 indicated clearly that there was no hidden reciprocity between gardeners' cash gifts to their husbands and their husbands' material support for the garden boom (see discussion in Schroeder 1993a). Nor was there any clear connection between substantial in-kind contributions by the gardeners (bags of rice) and male support. The data show that while nearly three-fourths of women in the sample either provided cash gifts or purchased supplemental grain to assist their husbands in meeting household requirements, very few men assisted their wives in any way with their gardens. Indeed, the support from men was lowest (86% of the sample provided neither labor assistance nor financial support to their wives) in households in which women made the highest contributions (when women gave their husbands both rice and cash gifts from their garden proceeds). Thus, there appears to have been little connection between women's economic contributions to the household and men's material support for gardening. The survey also ruled out the relative poverty of gardeners' husbands as an explanatory variable (see discussion in Schroeder 1993a). The only clear association was between a woman's personal wealth and her propensity for gift giving. Breaking the sample into thirds by income earned from gardening, I determined that 52 percent of the wealthiest gardeners gave their husbands cash gifts, nearly 80 percent supplemented family food stocks with a bag of rice, and over 90 percent gave either cash or rice. These findings lead to the conclusion that sizable material contributions by gardeners toward household well-being stemmed not from hidden reciprocities or the needs of impoverished husbands but from the desire of wealthier gardeners to encourage their husbands to relax control over their (the wives') labor; in short, gardeners used cash gifts and in-kind contributions to buy goodwill.

Obviously, not all women in the sample could afford lavish gifts for their husbands. Among the poorest third in my sample, only 8 percent of the women gave cash gifts, and 11 percent gave rice. The point here is not that all women were in a position to adopt such practices but that those who did were successful in swinging the moral economy of vegetable-growing communities in their favor. In this regard, the effect of the gift giving was quite decisive. Witness the following statements of two men married to Kerewan gardeners:

Today no one would say ["she's gone to her husband's"]. . . . Every man who is in this village whose wife is engaged in this garden work, the benefit of the produce goes to him first before the wife can even enjoy her share of it. That is why those statements they used to say would not be heard now. . . . In fact some men among us, if it were not for this garden work, their marriages would not last. Because their [own economic] efforts cannot carry one wife, much less two or three, or even four. Women can [now] support themselves. They will buy beds, mattresses, cupboards, rice . . . from the produce of these gardens. . . . In fact I can comfortably say that gardening generates a greater benefit than the groundnut crop that we [men] cultivate. Before you offer any help to people farming groundnuts, it is better you help people doing gardening, because we are using gardening to survive.

At the moment, a man cannot get from his groundnut farm what a woman can earn from her garden. Not even two bags of groundnuts in some cases. In the whole of the village, you can [easily] count the number of men who have eight bags [the rough equivalent of D1000]. And out of that [you must subtract] seed [and] fertilizer . . . it is only the women's sector that contributes greatly at this moment. When they are developed, the men will also develop.

The impression left by these comments is that the choices women made with regard to the disposition of their garden incomes, some motivated by compassion and others of a more strategic nature, met their mark. There is a third possible interpretation of these actions, however. According to some male informants, the disposition of women's garden incomes was not *purely* a matter of choice. They pointed out that men also actively pursued opportunities to gain access to their wives' money. In other words, the cash gifts and in-kind contributions women made to their families could be construed as a "taking" by men. This proposition requires closer inspection.

THE PRICE OF AUTONOMY

[Our husbands] thought we were wasting our time in the bush, but
when they realized the benefits they started praying for us.

Kerewan gardener

Open admissions by men in the Kerewan garden district that they con-
sciously engaged in maneuvers to gain access to their wives' incomes
were understandably quite rare. Those who did divulge information on
this topic stressed the difficulty of generalizing about the strategies they
were describing and felt it important to emphasize that they and their
male peers only engaged in such practices when they knew that their
wives could afford to share their assets. These caveats notwithstanding,
the data shed a great deal of light on the process of "negotiation" and
mutual "accommodation" precipitated by the boom.

The first set of strategies can loosely be described as loan seeking. It
consists of several different circumstances under which men asked their
wives for money, each with its own degree of commitment toward eventual repayment, and its own threat of reprisal if the funds were not forthcoming. The simplest scenario involved asking for a loan with no intention whatsoever of repayment. In this case, the crucial consideration for
the husband was how much to request. If he aimed too high, his request
might not be granted because his wife could legitimately say she did not
have the means. Also, if she did give him a larger sum, she would be
much more likely to either insist upon repayment or refuse to grant him
additional loans should he fail to make restitution. The ideal, then, was
to ask for a substantial amount in order to make the request (and its attendant loss of face) worthwhile, but to keep the request small enough
that the eventual financial loss could be absorbed or effectively written
off by the woman without retribution. Informants indicated that a request for D40–50—slightly less than an average week's net earnings—
would be a reasonable amount in most cases.

After defaulting more than once on repayment, or upon encountering resistance from his wife, a man might resort to the use of an intermediary who would request the loan on the husband's behalf. There
were actually two or three different scenarios in which this occurred. In
one, the woman in question realized that the third party was acting as
a surrogate for her husband. She nonetheless participated in the transaction willingly, since she knew that, in the event of default, she could
at least pursue the matter through the traditional court system. The use

of the courts was a step she would not consider taking in the case of her husband's direct default. The traditional elders who ran the courts would simply assume that the money involved in the loan transaction was used by the man for joint family benefit (whether in fact it was or not) and rule in his favor.

A second case of loan seeking via intermediary took place under conditions in which the wife was not aware that the loan was actually intended for her husband. This option was often chosen in situations when the husband had already exhausted his other more straightforward loan prospects, or in the event he was simply too ashamed to ask his wife for cash directly. From the husband's perspective, this approach retained the main advantage of the first form of "indirect" loan: his wife would more willingly acquiesce to terms under the assumption that third party loans were more enforceable than direct loans between marriage partners. Moreover, the husband's prestige would not be sacrificed in the process. In practice, however, he still had to meet the terms set by his surrogate for repayment.

A variation on the strategy of loan seeking via intermediary occurred when a third party, typically a junior family member, or even a child, approached the *husband* for a loan. In this situation, the man might choose to refer the would-be loan recipient to one of his wives. ("Presently if any child asks his / her father to buy anything for him / her, he will say to that child, 'go to your mother.'") It is worth noting that the question of whether or not the husband had cash of his own at the time of the request was moot. Indeed, if he *did* have cash on hand, his objective in diverting the loan request might revolve around that fact precisely: his aim was to protect his personal assets and shift the loan burden onto his wife's shoulders. It is also important to recognize that, under such circumstances, the weight of the moral economy shifted with the transfer of the loan obligation. Indeed, women were forced into a difficult position: if they refused to grant the loan request, they appeared hardhearted; if they pushed too hard for loan repayment, they strained family and friendship ties; if they chose not to pursue repayment at all, they forfeited their assets.

Two other more casual ploys rounded out the gamut of loan-seeking behaviors. Both entailed the regular battle between husbands and wives over everyday petty cash expenditures, or what I will refer to collectively as "fish money." These involved cash outlays for meat, fish, cooking oil, sugar, condiments, matches, candles, kerosene, flashlight batteries, laundry soap, in short, all the basic recurrent expenditures of everyday life in rural Gambia. Typically, the woman (or a small child sent on her behalf) mentioned to her husband as he was about to leave the family living quarters for the morning that she needed money to buy fish so she could cook lunch. This request was sometimes timed deliberately so that the exchange took place in front of guests, an added embarrassment for the husband. His response would be to complain that he had no money, and he would ask her to "help" him *(maakoi)* with a small loan. Alternatively, he might leave the house in the morning with the deliberate intent of shirking on the "fish money" obligation altogether before his wife even had a chance to ask him for money. In each of these cases, the net effect was the same—the wife ended up paying out of her resources for something that should, by custom, have been the husband's responsibility.

A woman's failure to provide a loan or pick up everyday expenses often resulted in a variety of sanctions being imposed upon her. An extreme response was for the husband to resort to outright theft. While my informants emphasized that this tactic was rare, they acknowledged that such incidents did occur. Much more common were the quarrels men initiated in order to raise the stakes in money matters. The basic strategy was for the husband to carefully select a pretext for picking a fight with his wife. It was considered ideal if the incident did not occur immediately after the loan request was denied; nor should it occur so long after the request that the connection was obscured altogether. My informants provided two hypothetical examples. In the first, the husband returned home unannounced from a firewood-cutting expedition, or a hard day of work on the family's fields. He arrived at a time when he knew his wife was either in her garden or had yet to draw the evening water supply from the town tap (in Kerewan in 1991, the public taps were routinely closed between 10 AM and 5 PM). He then demanded to know why there was no bath water waiting for him, complaining: "I came from the farm very tired and dirty, and this woman wouldn't even help me with bath water!" In the second scenario, he intervened as his wife administered a beating to one of his children for some obvious infraction: "How can you be so cruel to your daughter!" The development of these scenarios provoked a great deal of laughter among my informants. When asked why they found these stories so funny, they explained, "Because they are so *typical!*"

An actual incident that occurred in one North Bank community in

1991 involved a more extreme form of reprisal. A group of women described a domestic dispute that took place during the month of Ramadan, when practicing Muslims are expected to fast from sunrise to sundown. A row broke out at 6 AM when a man beat his wife, ostensibly because she failed to provide him with water to perform the predawn ablutions that mark the opening of the day's fast. The women recounting the incident roundly condemned the man because they were convinced that the real motive behind the beating was retribution after the man's wife refused to honor a loan request.

It is important to note that in all of these cases the husband's strategy was to try and occupy the moral high ground; even in the more ambiguous Ramadan case, the man could claim that his prayers were disrupted by his wife's failure to perform her "wifely duties." Moreover, it should be apparent that choosing a pretext for a fight in the context of the garden boom was a simple matter. With women routinely absent from family compounds and cutting corners in order to juggle competing demands on their labor, men were in a position to selectively invoke the abrogation of any number of traditional norms governing marriage relationships. The message, in any event, was quite clear: women who did not comply with requests for cash and acquiesce in the niceties of the loan-seeking charade paid a different sort of price. It did not take many beatings and shouting matches before this point hit home.

To be sure, the tactics men used to alienate garden income did not always poison social relations in this manner. There was in fact little evidence that the incidence of domestic violence rose as a consequence of the garden boom. If anything, it may have declined. My informants produced a short list of strategies under the general heading of "sweetness" (diya) that showed how in some cases the boom promoted marital ties that were at least superficially stronger and more cooperative. For diya entailed the husband being exceedingly nice to his wife—he might flatter her or "butter her up" with sweet talk. He might support her position in public discussion, or even advocate on her behalf on matters of substance having to do with her garden.[10] Alternatively, he might offer material support by (1) contributing labor; (2) lending her his donkey cart,[11] or (3) providing a small cash loan. He thus placed himself on secure footing with his wife in order to benefit from her good graces when she decided how to distribute her earnings after marketing her produce.

The final set of strategies employed by men seeking to control their wives' money entailed decisions over the disposition of their own cash-crop returns. I have already alluded to the fact that men routinely defaulted on the financial obligations they were expected to fulfill ("If you tell your husband to buy you a shirt or a pair of shoes, he will say you are crazy, I have more important things to do"). Much of this behavior could justifiably be attributed to the generalized economic hardship that accompanied the economic trends of the 1980s. Above and beyond such "legitimate" default circumstances, however, were steps taken by men to default on their responsibilities *deliberately*. This they accomplished by quickly disposing of their own cash assets before the exigencies of everyday life ("fish money," third party loans) absorbed them.

The key consideration for men in such circumstances was to choose an investment target that met with the tacit approval of his wife or wives. Examples of expenditures that would most likely be fully sanctioned include: the purchase of corrugated zinc pan or concrete for a construction project on the family living quarters, acquisition of a horse or donkey or additional farming equipment, and payment of costs associated with ceremonial occasions such as circumcisions or dependents' marriages. Likewise, investment in a seasonal petty trading venture would be largely beyond reproach on the grounds that some joint benefit could potentially be derived from the income generated by the husband's sales efforts. Far less welcome would be the purchase of luxury items such as a new radio, fancier furniture for the husband's personal living quarters, or expensive clothing. Money spent on other women might also be questioned, but that determination depended greatly on individual circumstances. A middle-aged woman without a co-wife might not object strongly to her husband marrying again since she would stand to benefit from sharing her domestic workload.[12] However, when the husband already had more than one wife, and his money from groundnut sales or salary payments provided little or no apparent joint benefit to his wives, his wives might well assume that he was squandering his money on gifts to girlfriends, perhaps the most "illegitimate" expenditure of all. This and other deliberate default practices were not only frowned upon by gardeners but were actively resisted, as the next section demonstrates.

BUYING POWER*

A Mandinka woman is still the same. It's the men who changed.
                                                North Bank gardener

*Thanks to Dorothy Hodgson for suggesting this section title.

The description of tactics I have compiled thus far establishes that women did not simply buy their husbands' goodwill outright. Men asserted their advantages wherever they could to shift the balance of economic power in the household (back) in their favor. Husbands were not in a position to leverage their wives' consumption choices at will, however. Indeed, there is considerable evidence that women were firmly resolved to protect their interests, as the following quote demonstrates:

> Our husbands stopped buying soap, oil, rice. . . . We provide all these
> things. Obviously our marriages would change. We do all this work
> while our husbands lie around home doing nothing. Whenever we
> return from gardening, we still have to do all the cooking, and all our
> husbands can say is, "Isn't dinner ready yet?" And then they start to
> shout at us. Remember, this is after we have already spent the whole
> day at the garden working. . . .
>   A husband who has nothing to give to his wife—if that wife gets
> something from her own labor, she will surely find it more difficult
> to listen to him. We women are only afraid of God the Almighty.
> Otherwise we wouldn't marry men at all. We would have left them
> by themselves. . . . Men are always instructing us, you better do this or
> that for me, while they sit at the *bantaba* [the neighborhood meeting
> place] all day doing nothing. They describe us as foolish, but we are
> not, and we will not listen to them.   *North Bank gardener*

Women used several different strategies to protect their cash incomes. The most basic approach for a woman was to prevent her husband from ever knowing how much cash she had on hand in the first place. This required that she adopt a "false face" of sorts within the family compound (Pred 1990; Scott 1990), as though she were not engaged in a complex year-round production system involving perhaps a dozen different crops, grown in three or four sometimes far-flung locations, each generating its own seasonal pattern of income. In order to create and maintain this fiction, women rarely discussed garden matters with, or in the presence of, their husbands. This resolute silence stood in sharp contrast to the running discussion and debates women engaged in along the footpaths to and from, and in, the gardens themselves. A veritable stream of information concerning prices available at the different North Bank market outlets *(lumoolu)* was exchanged as women moved about and tended to their crops.

Many gardeners hid their income through use of intermediaries to carry produce to market on their behalf. Survey results showed that well over half of the women in my research sample relied at least occasionally on someone else to carry produce to market for them (see discussion in chap. 4). Others shipped produce to market directly from garden sites. In this way, their husbands were prevented from actually seeing the produce assembled in one place, an opportunity that might give them a clearer sense of how much their wives actually earned. Women also sequestered their savings in such a way that they could not be touched by their husbands. This they accomplished in a literal sense by wearing "money belts" on a regular basis. With respect to larger cash sums, they often gave their money to older female relatives or trusted neighbors for management and safekeeping. In one village, for example, gardeners opened up savings accounts with a local shopkeeper (cf. Shipton 1995). Parallel records were kept by the shopkeeper and a trusted local civil servant indicating running balances on individual accounts. Assets were thus protected from seizure by the merchant, who was held accountable by the civil servant. At the same time, the shopkeeper paid no interest and was free to use the cash to capitalize his business or engage in money-lending. In exchange, women benefited from keeping their assets relatively liquid without exposing the extent of their accumulation to their husbands directly.

Even with such diversionary tactics in force, the peak of the marketing season almost inevitably brought with it increased "loan-seeking" behavior on the part of men. Consequently, the second major area of attention for women concerned controlling the terms under which loan agreements were undertaken. Thus, if a woman's husband repeatedly defaulted on loan payments she might choose to stop granting him loans altogether. Alternatively, she could wait for, or insist upon, the intervention of a third party to the loan transaction. In the relatively rare event that this failed to generate the desired outcome of a reasonable repayment rate, the woman might choose to go the risky route of public disclosure. Airing the dirty laundry of intra-marital finances was a virtual invitation to divorce; the messiness of such a scandal would almost certainly damage the woman's reputation along with her husband's. Such a course was, nonetheless, sometimes preferable to enduring the repeated predatory demands of a greedy husband. Yet another strategy involved the woman pre-empting her husband's loan requests by giving him cash gifts before he even asked for them. Such gifts were an attempt to carry out an increasingly obligatory transfer of assets under terms that the women themselves could control: rather than suffer their husbands' whims, women determined both the amount and the timing of their gifts, thus inoculating themselves against unexpected and exorbitant loan requests that might otherwise disrupt personal plans at inopportune moments.

Finally, when all else failed, women simply opted for the same solution widely employed by their husbands—they tied up their cash assets by spending them as quickly as they received them:

What happens is, some men would like their wives to loan them some money out of their garden sales. Many times women will grant the requests, but most of them will never be refunded. So women gradually limit, or refuse, credit to their husbands. We have a new tactic: when we go to market [to sell our produce], we simply spend all our earnings on things that we need, and come home with no money at all to avoid the loan requests altogether.   *North Bank gardener*

Among the items women might buy under such circumstances were dowry items for their daughters such as dishes, pieces of cloth, or jewelry. While some of the men interviewed bitterly criticized their wives for assembling overly lavish trousseaus for their daughters, this tactic was sometimes simply a woman's response to her husband's own profligate spending habits. In cases where deliberate default was mutual, the family's financial security was obviously placed in jeopardy, and the marriage itself rested on quite shaky ground.

GONE TO THEIR SECOND HUSBANDS

It is fair to say that domestic budgetary battles did not originate with the garden boom in Mandinka society; nor are they wholly unique to either The Gambia or Africa (Guyer 1988; Whitehead 1981). Nonetheless, the Gambian garden boom clearly produced dramatic changes in the normative expectations and practices of marital partners in the country's garden districts. In the context of climate change, new foreign investment patterns, and structural economic adjustment, the growth of a female cash-crop sector virtually inverted the economic fortunes of men and women in many parts of rural Gambia. Discursive politics played a prominent role in the negotiations that accompanied those changes. The wielding of marital metaphors as weapons in a battle to seize and / or regain the moral high ground resulted in something of a standoff: men used the claim that women had become "married" to their gardens to force their wives to transfer control of at least a portion of their assets or face continued verbal (and on rare occasion, physical) assault. Women gardeners appropriated the metaphor to underscore the repeated failure of their husbands to provide for their families. In doing so they claimed for themselves the freedom of movement they required to go about their gardening tasks unimpeded. In short, the metaphor equating gardens with husbands encapsulated the mutual default of *both* marriage partners with respect to customary responsibilities.

Generalizing on societies with "a pronounced division into male and female spheres," Jane Guyer notes that "the specialization [of budgetary responsibilities] is never complete; it oscillates according to each sex's ability to cope with its own sphere, and its ability to tap into the other or to shift the responsibilities" (Guyer 1988, 171–172). The "ability to cope" in rural Gambia was directly tied to the capacity of individuals to earn cash incomes, and thereby to the respective fortunes of the separate crop production systems. These fortunes varied widely by household; they also hinged on factors such as climate and international market perturbations that were well beyond local control. By contrast, the "ability to tap into" another sphere or "shift responsibilities" was directly related to the localized power dynamics that took shape in the garden districts. These had to do with moral economic forces, strategies of deception, and the tactics of marital negotiation concerning property, income, and power relations.

Since "coping" strategies and ruses designed to shift responsibilities were in play at all times during the early years of the garden boom, it is extremely difficult, from an analytical standpoint, to prise the two apart. Negotiations over cash transfers between men and women became— quite literally—give-and-take situations. Loans were loans until men stopped paying them. Then they either became "cash gifts," as described above, or the source of more serious struggles leading to divorce. By a similar token, gardens were "husbands" that controlled women's labor, until they became "husbands" providing food for women's families. Such ambiguity inflects a final reading of women's "autonomy" in the context of the garden boom. While there was evidence that acts of accommodation undertaken by women on the North Bank softened the rhetorical stance their husbands once took against gardening, this achievement did not alleviate the pressure on women entirely. They were still required to meet a rigorous set of financial obligations: not only were they forced to contend with the domestic financial squeeze engineered by their husbands, but they did so under the pressure of surplus extraction from traders, produce transporters, and landholders (about which more below). This double bind was exacerbated in drought years, when the irrigated vegetable crop became one of the only bastions against generalized food shortage and extreme economic hardship.

The price of autonomy notwithstanding, women in The Gambia's garden districts succeeded in producing a striking new social landscape—by embracing the challenges of the garden boom, they placed themselves in a position to carefully extricate themselves from some of the more onerous demands of marital obligations. Indeed, in a very real sense, they won for themselves "second husbands" by rewriting the rules governing the conjugal contract. Thus the product of lengthy intra-household negotiations brought on by the garden boom was not the simple reproduction of patriarchal privilege and prestige; it was instead a new, carefully crafted autonomy that carried with it obligations and considerable social freedoms.

Better Homes and Gardens

*The Social Relations
of Vegetable Production*

The new conjugal contract won by gardeners on The Gambia's North Bank accorded them considerably greater autonomy than they had held previously, but only on condition that they continue to meet their husbands' demands for financial support. Thus in the early stages, the intra-household battle over domestic budgets became a major driving force behind the garden boom. Gardeners responded to this impetus by pressing hard to intensify production, even while continuing to satisfy their families' needs for domestic services. The primary objective in this effort was to extend the income stream from gardens beyond the narrow market window afforded by crops traditionally grown in the area. In order to meet this challenge, gardeners had to confront several key constraints. First, they were forced to locate and develop markets for their fresh produce, a notoriously difficult proposition for vegetable growers the world over. Second, they had to overcome seasonal barriers to vegetable production. Water supply management, crop protection measures, and pest and disease controls were all made much more difficult by extending vegetable production beyond the early dry season. And third, they had to develop a system for coordinating a multi-generational labor pool that could successfully meet the demands of their complex, multi-crop production regimes. This task was complicated by sometimes harsh local growing conditions and the spatial organization of the garden perimeters.

In short, gardeners working to extend the North Bank's garden boom engaged the task of *calibration* (Berry 1987; Carney and Watts 1991). Forced by circumstances to make dozens of adjustments in marketing and work routines, they struggled to bring about the mutual accommodation of their system of intensified vegetable production, the pre-existing farming system, and the system of domestic reproduction, all of which relied heavily on female labor. The summary impression is one of a tight production schedule reinforced on all sides by intense social and economic pressures.

THE SOCIAL RELATIONS OF VEGETABLE MARKETING

The Wolof pun on the name "Kerewan" that dubbed the civil service outpost *Kerr Waaru*, or "place of frustration," had special significance in light of the market situation facing the village's gardeners. Although the area under cultivation by Kerewan's growers had increased six-fold in twenty years, with average incomes from fruit and vegetable sales increasing proportionately, these gains were hard fought. On the one hand, Jowara Creek, immediately to the west of Kerewan, was un-bridged (map 3). An unreliable ferry service slowed traffic on the North Bank road and thereby limited growers' access to relatively lucrative Banjul-area markets. On the other hand, the international border presented an obstacle to more extensive trade with the major Senegalese centers of demand in Kaolack and Dakar, each within a day's journey to the north. The question, then, is how Kerewan women managed to sell their produce at all.

The marketing options Kerewan women possessed in 1991 were actually several, and gardeners' preferences varied over the course of the trade season. In order of increasing distance from the gardens themselves (map 2), the choices can be classified as (1) *local:* traders *(juloolu)* who bought vegetables at the garden gate, sales within the village itself, sales at the Kerewan ferry terminal, and hawking *(kankulaaroo)* in Darsilami and other nearby Jokadu District villages; (2) *intermediate:* the Kerr Pateh, Kerr Musa Saine, and Ndungu Kebbe weekly markets *(lumoo,* sing.; *lumoolu,* pl.*)*; and (3) *long-distance:* the Farafenye *lumoo,* Barra market, and the greater Banjul area. Transportation costs played a major role in determining which options women actually chose (table 6). Early in the season, conditions favored the use of local options. Economies of scale did not yet exist for the transportation of produce to more

TABLE 6   ECONOMIES OF SCALE
IN VEGETABLE MARKETING

Outlet	Transport Costs[a]		Price	Net Return per Pan Shipped[b]		
	Per seller	Per pan of produce	Per pan cabbage in peak season	1 pan	2 pans	3 pans
Darsilami	10	2	variable	n.a.	n.a.	n.a.
Kerr Pateh	14	3	30	13	20	22
Farafenye	30	6	40	4	19	24

SOURCE: R. Schroeder, field notes.
All figures in 1991 dalasis.
[a] The estimates of transportation costs reflect round-trip fares for the seller and one-way charges for produce.
[b] These calculations consider transport costs only.

distant outlets despite the premium prices available. More importantly, the opportunity costs of traveling to market were high at this time of year given the requirements of standing crops. Thus, at this stage, women often relied on their younger daughters to hawk their vegetables door to door or sell them to passing travelers at the ferry terminal. As yields increased, economies of scale dictated that growers could profitably shift to the Kerr Pateh outlet, and a minority of growers began traveling as far as Farafenye or Barra. At the season's midpoint, wholesale vegetable supplies to Kerr Pateh market reached saturation level. Significant price differentials for cabbage and bitter tomato (*jakatu*) developed between the Kerr Pateh and Farafenye markets, and many growers shifted temporarily to the long-distance option.[1] Finally, as dry season production wound down in May and June, and both the need for a rest period and the ensuing preparations for a new rice crop asserted themselves, women returned to more local options for selling their produce and maintained that strategy throughout the rains.

The situation of Kerewan market gardeners seems to lend itself to an analysis based on time, distance to market, and weight and perishability of produce, in short, to the principles of neoclassical economic geography.[2] On closer inspection, however, a number of social and cultural factors had equal weight in shaping North Bank marketing practices. Two groups of women encountered special difficulties in marketing vegetables, and their situations help illustrate the point. Younger women, whose sexuality was more tightly controlled by their families (or who were pregnant or breastfeeding), were less likely to be granted leave to travel longer distances to trade than their more mature counterparts. In part, this was due to the fact that trading in Farafenye, the long-distance option of choice for most women, entailed traveling the night before market day and sleeping on mats in the open at the marketplace. This practice was considered unseemly for young women and impractical for mothers caring for newborns. On the other hand, elderly women who were freer to move about as they chose were not always able to withstand the rigors of transporting bulk quantities of goods on long journeys. They would therefore confine their trading activities to local options. In both of these cases, the frequency with which a woman's goods reached outside markets hinged on whether a surrogate could be arranged to transport the produce on her behalf.

Economies of scale would seem to dictate the need for cooperative solutions to these marketing dilemmas. However, women in Kerewan engaged in reciprocal marketing efforts involving non-family members only occasionally as special needs arose (table 7). As one woman explained, the risk of social conflict was too great to send a non-family surrogate to market. Prices were volatile enough that two women traveling to the same market on the same day with the same complement of vegetables might receive very different prices for their goods. Since there was no way of verifying whether or not a surrogate had bargained in good faith on one's behalf, women typically relied on family members with whom they shared at least some financial interests in common to carry their vegetables to market for them.

Cooperative relations did play some role in shaping North Bank marketing practices, however. Although Kerewan was the first village to move decisively in the direction of intensified vegetable production in the 1970s, it was later joined by several other Lower Baddibu District villages with access to low-lying land. This concentration of garden villages accounts for the serious market gluts and low producer prices Kerewan growers experienced at peak season at the Kerr Pateh *lumoo*. The price dynamics were not purely mechanical, however. In 1989, for instance, prices were exceptionally low for most vegetable commodities, and growers were convinced that some form of collusion on the part of traders was taking place. In Kerewan, growers widely condemned these alleged practices, and discussion and debate quickly centered around the prospect of mounting a market boycott. Fearful of the reception such an

TABLE 7   MARKETING OPTIONS USED
BY KEREWAN GARDENERS, DRY SEASON, 1991

Preferred outlet	Local	15
	Intermediate	64
	Long-distance	20
Frequency of sales	Steady, every week	41
	Every other week	31
	Variable	23
	Sell to meet demand	4
Marketing agent	Self or member of work unit	48
	Family member with own plot	29
	Occasional help from nonfamily	18
	Nonfamily reciprocal arrangement	4

SOURCE: R. Schroeder, survey data.
NOTE: $n = 99$.

unprecedented action might get from local authorities, the leaders of the boycott initiative sent a delegation to the district chief *(Seyfo)* of Lower Baddibu District in the nearby village of Saba to explain their position. After hearing the Kerewan delegation's complaint, the Seyfo endorsed their plan and took the additional step of calling together representatives of all the major gardening communities within his jurisdiction to more carefully coordinate the action.

At the ensuing meeting, delegates from five communities agreed to stay away from the *lumoo* indefinitely in an attempt to force traders out into the villages to conduct their business at the garden gate. In theory, this arrangement would have favored growers in several respects. The growers would have saved transportation costs (which accounted for roughly 20 percent of the value of gross sales in Kerewan in 1991), avoided the heavy losses that sometimes occur due to spoilage during shipping (Holcomb 1991) and minimized time lost in transit.[3] The sellers' bargaining position in price negotiations would have also been strengthened considerably, since the onus would be placed on the buyers to make their investment in transportation pay off. Indeed, had the boycott come off as planned, growers throughout Lower Baddibu District would have been in a position to simply harvest to meet demand.

In the end, the outcome of the boycott strategy was mixed. On the one hand, traders did venture out of the *lumoolu* to buy produce in villages once the boycott was in force. On the other, their forays into the growing districts failed to benefit all villages equally. Kerewan women missed out on the benefits of the boycott because women in other villages closer to Kerr Pateh pressed their comparative locational advantage to capture most of the decentralized trade. In less than two weeks, the same Kerewan growers who organized the boycott in the first place broke ranks and returned to the market to salvage what they could from an increasingly desperate situation.

As the 1989 market boycott suggests, the difficulties women experienced in marketing their vegetables were one of their greatest sources of frustration. This fact was painfully evident when I asked a group of women to compare the problems they experienced in marketing their vegetables with those their husbands encountered in the sale of groundnuts. My purpose in drawing the comparison was to try and sort out the reasons women consistently problematized their situation in terms of gender rather than class. For the sake of argument, I noted that prices for groundnuts over the previous decade had been generally poor and that the various market outlets used by groundnut farmers had been prone to different forms of corruption and manipulation. I pointed out that both men and women were in relatively weak bargaining positions when it came to marketing, and this, in theory at least, would seem to afford some basis for common ground between the two groups.

The vegetable growers I spoke with immediately scorned the crudeness of my comparison. They cited linguistic barriers,[4] a lack of familiarity with Senegalese currency, and limits on mobility due to the constraints of domestic work as evidence that women were at a greater disadvantage in marketing produce than men. They also pointed out the example of a nearby village where men had become directly involved in the marketing of vegetables for their wives. The Kerewan growers noted how these men routinely received higher prices in the market than the Kerewan growers themselves received. When asked why men in Kerewan did not help their wives in the same manner, one woman brought the discussion to a resolute conclusion with the deeply bitter reply, "We have no men here."

As table 7 shows, then, decisions on who went to market, how often, and which market was used covered a wide range of options. Individual strategies were determined jointly on the basis of economic, social, and cultural criteria, as well as more straightforward economic geographical considerations. Returns to labor and the success with which women could continue to purchase their freedom to garden from their husbands varied accordingly.

SEASONALITY AND GARDEN INCOME

The lack of male support in marketing efforts and the efforts by men to alienate their wives' income detailed in Chapter 3 combined with the effects of drought and structural adjustment to force gardeners to intensify their production. In the mid-1970s, as the garden boom was still taking shape, the Gambian government selected several communities with sizable "kitchen gardens" as the target sites for an extension program directed at the mono-cropping of onions. This promotion, a modest governmental attempt to reduce local dependence on food imports (Mickelwait et al. 1976) achieved some early success but quickly fell out of favor with growers in Kerewan. Onion sales were repeatedly undermined by market gluts, and relatively poor market returns placed women in the untenable position of being absent from family compounds for long hours without significantly enhancing their incomes in return.

Consequently, Kerewan's gardeners abandoned their concentration on onions, and diversified their crop selection:

Fifteen to twenty years ago, you would not see anybody who would harvest 50 kg. of bitter tomato, fresh tomato, and so on, just to mention a few [of the crops currently grown]. But now we produce these vegetables by the 100 kg. bag. Garden work started here with spring onions [shallots] and small hot peppers. From there we had additional vegetables like bitter tomato, okra and small fresh tomato variety. But now you can see for yourself that the variety of vegetables we grow in our gardens includes eggplants, cabbages, onions, big fresh tomatoes, big and small chili pepper, carrots, lettuce, bitter tomatoes, and others.   *Kerewan gardener*

The importance of crop diversification strategies lay in the fact that they allowed gardeners to prolong the market season. As the data in table 8 indicate, Kerewan's leading crops in 1991 had switched from the onions (8% of total sales), tomatoes (8%), and chili peppers *(kani mesengo;* 1%) that dominated production in the 1970s to cabbage (39%), bitter tomatoes or aubergines (26%), and capsicum peppers *(kani baa;* 20%). By broadening their crop selection, growers not only minimized market risks, they also created an income stream that was more evenly distributed and therefore less apt to be diverted by husbands' loan requests.

TABLE 8   VEGETABLES GROWN AND SEASONALITY OF
INCOME, KEREWAN, 1991
(sales estimates, dalasis per week)

	Aubergines	Cabbage	Onions	Capsicum Peppers	Chili Peppers	Tomatoes	Total
15 February	100	280	0	0	0	75	455
22 February	900	795	135	0	0	225	2055
1 March	1400	1395	0	0	0	480	3275
8 March	2625	2145	270	0	0	840	5880
15 March	3575	3525	120	35	0	1350	8605
22 March	3625	3705	180	105	0	1590	9205
29 March	3375	5895	1125	480	0	1245	12,135
5 April	1775	3975	1238	1020	0	1155	9193
12 April	2280	3705	938	2400	50	555	9958
19 April	975	3140	1125	2580	30	460	8340
26 April	1800	3712	1100	4235	150	285	11,353
3 May	1260	3105	875	2190	100	45	7593
10 May	930	2150	420	2680	200	30	6428
17 May	800	1550	380	2520	60	15	5283
24 May	760	1300	100	1400	60	0	3595
31 May	1240	1500	455	1260	0	0	4455
7 June	520	420	0	140	30	0	1128
14 June	100	255	0	340	0	0	713
Total	28,040	42,552	8460	21,385	680	8350	109,645
% total	25.6	38.8	7.7	19.5	0.6	7.6	

SOURCE: R. Schroeder, survey data.
NOTE: $n = 100$.

In the early stages of the boom in Kerewan, before the rapid expansion of garden enclaves, the only way diversification could be accomplished was by intercropping or double-cropping. Predictably, these strategies were common among Kerewan growers. Of a sample of 100 growers surveyed in 1991, only 5 percent adhered to strict mono-cropping principles in all of their garden beds, while the remainder practiced some form of intercropping. Forty-seven percent of Kerewan gardeners followed double-cropping practices (i.e., the initial crop was harvested before the second crop was planted). And 73 percent of the Kerewan sample practiced serial or relay cropping. This approach entailed planting companion crops in the same bed that matured at different rates and were harvested and marketed at different times. Growers frequently planted cabbages and capsicum peppers in series, for example. Cabbages were harvested first, and the productivity of the late season pepper crop was then enhanced through timely doses of fertilizer.

The benefits of seasonal expansion notwithstanding, the longer growing season posed serious difficulties for gardeners. First, they had to confront the problems of maintaining a secure dry season water supply,
without which gardening would be impossible. As table 5 indicates, the
recurrent expenditures associated with irrigation were significant, although the precise nature of the problem differed by location. In gardens
in low-lying swampy areas with a water table as shallow as Kerewan's
(2–5 meters), many of the wells were hand dug, either by local welldiggers or by the gardeners themselves. While outside technical support
was not required for such tasks, the cost of well maintenance to individual gardeners was high. Since most hand-dug wells were unlined,
they were subject to frequent collapse and thus required frequent deepening or redigging. Survey results showed that 100 Kerewan gardeners
dug 103 new wells and deepened or repaired 192 existing wells in 1991
due to partial collapse and seasonal fluctuations of the water table.[5] The
maintenance and replacement of wells, which accounted for 13 percent
of garden costs in Kerewan in 1991 (excluding labor and land costs),
constituted a fairly constant, recurrent expense for gardeners. The principal obstacles to maintaining a steady water supply were thus expense,
and / or finding welldiggers capable and willing to perform the necessary
tasks under tight seasonal constraints.[6]

A second major problem gardeners faced had to do with crop protection. During the rainy season, all livestock owned by residents of Gambian towns and villages were corralled, tethered, or otherwise herded
under close supervision in order to avoid damage to standing crops in
fields and swamplands.[7] By contrast, in the dry season, most livestock
were allowed to range freely so they could glean whatever meager forage resources remained on the ground following the harvest.[8] As the
dry season wore on and these resources were exhausted, the pressure
by goats, sheep, cattle, and donkeys on the relatively lush fenced garden
perimeters increased steadily. Prior to the garden boom, most gardeners would simply abandon their plots after the short market season, and
the animals would be allowed to graze unimpeded. After the opening
of commercial outlets for vegetable sales, however, much more was at
stake. The importation of barbed wire and chicken wire for fencing purposes alleviated some of the problem, but routine fence maintenance
and the nearly constant physical presence of women in the garden during daylight hours were both required if intruding livestock were to be
successfully kept at bay.

In this regard, the fencing problem intersected with a third problem domain centered on plant pests and disease. As gardeners pushed their production schedules into May, June, and July, insect pest and disease problems increased. Plants weakened by drought stress under conditions of late dry season heat succumbed much more readily to disease. Moreover, during these months insect pests and birds had few alternative hosts, and their assault on the gardens intensified. The effect of these problems on gardeners was cumulative. As pest and disease problems mounted, there was a gradual decline in the number of growers maintaining an active presence in the garden. As the dry season drew to a close and fewer women were actually physically present in the gardens on a day-to-day basis, it became more difficult to keep goats out of the perimeters, grazing damage to gardens increased, more women abandoned their gardens, and the downward spiral continued, leaving a small percentage of active growers until the rains began and livestock were once again tethered (Schroeder 1991c).

GARDENING AND DOMESTIC LABOR

The move toward an extended cropping season, evidenced by expanded crop production and the opening of new market outlets, was intended to resolve some of the vegetable growers' domestic budgeting problems but carried with it a new set of difficulties at the point of production itself. While these "technical" dynamics all had specific social and economic implications for men and women struggling over household budgets, they also had great significance for social relations *between* women, and the ways in which they organized their labor regimes.

The daily round of domestic duties performed by women in The Gambia was extensively documented by Britain's Dunn Nutrition Unit, which conducted its research under the institutional auspices of the British Medical Research Council (MRC). Details of MRC research, originally published in Roberts et al. (1982) and summarized in Barrett and Browne (1989), are included in tables 9 and 10. Table 9 shows both the range and level of difficulty of labor tasks shouldered by rural women; table 10 provides details of the time spent on selected domestic tasks, including a breakdown of the relative workload each task entails (compare Nath 1985a). These data demonstrate clearly the heavy demand for female labor in The Gambia. What they do not reveal, however, is how those demands increased following the garden boom, and in particular how the level of labor demand was influenced by seasonal growing conditions.

TABLE 9   WORK LEVELS
OF WOMEN'S ACTIVITIES

Hard Work	Moderate Work	Light Work
chopping wood	drawing water	shelling groundnuts
clearing land	laundering clothes	breast feeding
planting seeds	carrying light loads	
drawing large buckets	removing husks from	
of water from well	rice by pounding	
removing bran from	removing cereals from	
cereals by pounding	stem by pounding	
carrying heavy loads	sweeping and cleaning	
(greater than 20 kg)	walking to / from fields	
	guarding fields	
	collecting wood	
	cooking	
	child care	
	personal hygiene	
	shopping	
	weeding rice	
	transplanting rice	
	harvesting rice	

SOURCE: Roberts et al. 1982; as cited in Barrett and Browne 1989, 5.
These data were collected in Keneba, located in Kiang West District on the South Bank.

The excessive heat and aridity during the period from February to April when gardens bear at peak capacity contribute toward exceedingly high rates of evapotranspiration.[9] Successful horticultural production hinged, therefore, on the ability to carry out both early morning and evening watering rounds, since it is during these periods that plants can take up moisture most efficiently under intense evapotranspiration pressure. In a situation such as Kerewan's, where women held an average of three garden plots, often within more than one of the widely dispersed garden perimeters that flanked the village at a distance of up to 1.5 km, two trips to water gardens could easily require three to six hours away from family living quarters each day. Some women acknowledged the fact that they needed to cut corners to compensate for the multiple demands these circumstances placed on their labor: "As you know, the distance between home and the garden is a long walk. In order to perform garden tasks in a timely way, one needs to forgo some of the domestic chores at home." Others described how they deployed alternative labor sources to help manage their workloads.

The women in the Kerewan sample engaged in a variety of formal and informal cooperative labor relationships that alleviated some of

TABLE 10   TIME AND ENERGY EXPENDED
ON SELECTED WOMEN'S WORK TASKS
(time in minutes)

Activity	Total time	Workload		
		Hard	Moderate	Light
Cooking				
Boiled staple and sauce	94	2	46	46
Breakfast	16.5	0	14	2.5
Pounding				
Millet or sorghum off stem	117	43	40	34
Millet or sorghum, bran removed	44	25	8	11
Milled or sorghum, endosperm	66	28	10	28
Rice, husk and bran removed	51	26	23	12
Drawing water from well				
Two large buckets	49	7	22	20
Sweeping				
Compound	32	0	31	1
House and kitchen	10	0	10	0
Washing				
Family clothes at well	180	0	161	19
Bowls and pots for one meal	13	0	13	0

SOURCE: Roberts et al. 1982, as adapted in Barrett and Browne 1989, 5.

the social pressures outlined above. While 53 percent of growers surveyed reported that they worked alone on their crops, 37 percent were part of mother–daughter(s) work units, and the remaining 10 percent worked regularly with sisters, nieces, granddaughters, daughters-in-law, co-wives, or sons (table 11). The sharing of work responsibilities in such relationships depended upon the age, health, and fertility status of the women involved. Women who were in the latter stages of pregnancy or nursing newborns could be expected to reduce their workloads for a season or two. Elderly women, women who suffered from ill health, and young girls also had fewer responsibilities for day-to-day garden maintenance but were frequently enlisted to help with specific garden tasks.

An individual woman's entitlement to such labor services, where they were available, was largely contingent upon her position in the age hierarchy (compare Jackson 1995). A younger woman with no co-wife and

TABLE II   PRODUCTION UNITS
IN KEREWAN GARDENS

Normal Work Unit	Units in Category
Self only	53
Mother / daughter(s)	37
Self and other relative (5: sister, 2: niece, 1: granddaughter, 1: co-wife, 1: daughter-in-law, 1: son)	11
Total	101[a]

SOURCE: R. Schroeder, survey data.
NOTE: $n = 99$.
[a] Three women in the sample worked with more than one other person (e.g., with both their mother and daughter; no information was available for one woman). Half of the women reported additional occasional, informal cooperation between themselves and women in adjoining plots. Interestingly, the kafo, or age set, social structure so prominent in other ethnographies of Mandinka rice production (Carney 1986; Dey n.d) played little role in Kerewan's gardens.

no children old enough to assume supportive labor responsibilities was extremely limited in her options. If there were older women in her compound (e.g., her husband's mother), it was much more likely that they would rely on the younger woman for labor support than vice versa.[10] Moreover, a younger woman was more likely to be encumbered by the demands of her fertility career than her older peers. By contrast, older women typically benefited more from the productivity of their children. They were more likely to have co-wives with whom they could share certain responsibilities on a rotational basis. And their claims on the labor of kinswomen, both those older than themselves who no longer maintained their own plots and those younger, were far more extensive. This increased their options for delegating child care and other duties.

Exactly how a woman deployed the resources available to her was a separate question, however. There was almost no escaping the fact that the head of a gardening unit had to personally spend a great deal of time away from home if her gardens were to thrive, but there were different means of minimizing the social costs of this work routine. Two main work patterns emerged. In the "single visit" regimen approximated in table 12, women were continuously absent from the family compound from mid-morning until early evening every day. This pattern minimized the time lost in transit to and from the village and was thus best suited to situations in which the garden site was distant. It was not, however, a preferred strategy from the standpoint of gardeners' husbands. In at least one North Bank village, the routine, all-day absence of women from town for the duration of the dry season prompted a group of gar-

TABLE 12   "SINGLE VISIT" WORK REGIMEN
OF GAMBIAN WOMEN FARMERS

Time	Activity
05:00	wake, wash, dress
	collect water
	prepare and give children bath
	prepare lunch
	do laundry
	sweep compound
	eave children in compound with their lunch
10:00	leave for fields with lunch
11:00	start work in fields
	During the day, breaks are taken to breastfeed babies, to eat lunch, and to rest.
17:00	collect fuelwood and other bush products
18:00	leave fields to return to village
19:00	prepare evening meal
20:00	eat evening meal, wash up, and bathe children
21:00	bathe and retire for the night

SOURCE: Barrett and Browne 1989, 5.

deners' husbands to lobby the town chief to implement a ban on gardening outside village limits (Schroeder and Watts 1991).

The "double visit" work regimen, by contrast, rested on a more well-articulated division of labor which simultaneously took into account the strength and skill requirements of specific work tasks, the social significance of particular symbolic actions, and the spatial extent of the home- and garden-based work domains. In Kerewan, where most growers followed the "double visit" routine, a pattern of shift work was implemented. First thing in the morning, as their mothers performed the range of early morning domestic duties described in table 12, young girls aged six to twelve were sent to the garden to begin watering garden beds. This unskilled "moderately" heavy work task (table 9) was well within the girls' physical capacity, provided certain conditions were met. Most critically, wells had to be available in sufficient number and within close enough proximity to garden beds that the girls were not forced to carry water over long distances. This accounted in part for the remarkable density of wells in Kerewan's gardens. A physical inventory of roughly nineteen hectares of well-established garden plots indicated 1,368 wells in use.[11] This meant that, on average, each well served 139 square meters; the girls accordingly had to walk only an average of 12–14 m before encountering an area served by an adjacent well.

Figure 5. Reconfiguring domestic space. In order to accommodate the space–
time limitations of home- and garden-based work routines, North Bank
gardeners had to reconfigure domestic space in their gardens. In this context,
heavy imported plastic basins served many purposes. Primarily used to haul
water and fresh produce, they sometimes doubled as bassinets and laundry
baskets. As women went about their daily tasks, young children, aged five to
six, were often assigned to keep a watchful eye on their infant siblings.

At about eight-thirty, the young girls would finish their shift of irrigating crops and return from the gardens for breakfast, following which
they would receive instructions for a second set of tasks. These might
include drawing water, some of the less heavy grain pounding chores,
child care, shopping, food preparation or cooking (low-skilled, light, or
moderately heavy jobs that could be carried out either independently or
under the supervision of elderly relatives). While the girls were engaged
in these duties, their mothers hurried off to the gardens for two or three
hours to complete whatever watering remained undone and to carry out
the more discretionary aspects of cultivation, such as seed bed preparation, transplanting, and the application of manures, fertilizers, and pest
control measures. At noon, the women returned home to finish prepar-
ing lunch, personally tending to the more skilled task of sauce preparation and the deferential act of serving the dish to their husbands. In so
doing, they sought to avoid unnecessary confrontations over their absence from the compound. Then, between four and five o'clock in the afternoon, when the most extreme heat of the day had passed, women re-

turned to the gardens to carry out a second round of irrigation and at-
tend to a variety of other special tasks before dusk.[12]

The "double visit" irrigation system illustrates clearly how women
calibrated their work regimes to simultaneously address the social and
ecological constraints they faced on a daily basis in their garden production systems. There were also a number of variations on the single
and double visit approaches that invoked other sets of social relations.
Polygamous marriage arrangements gave women somewhat more flexibility in managing their gardens than did nuclear families. On "off"
days, when a married woman's co-wife bore principal responsibility for
meeting her husband's personal needs, she was likely to organize her
workday along the single visit model. Alternatively, when women were
the sole wives of their husbands, they frequently took both small children and domestic work with them to the gardens. Children as young as
three to five years were often pressed into service to take charge of infants and toddlers while their mothers tended to crops, did laundry, or
attended to other tasks in the relative cool of the garden (fig. 5).[13] Other
steps individual women took in the mutual accommodation of their
daily garden and domestic labor routines included the spatial consoli-
dation of landholdings to minimize time spent in transit between different garden sites and relocation closer to relatives who could help watch
children or tend crops when the women themselves were sick, in the advanced stages of pregnancy, gone on marketing trips, or otherwise unable to work a normal schedule. Finally, women occasionally defaulted
on domestic duties altogether. It was not uncommon, for example, to
find fathers tending children for a few hours in the late afternoon, as
their wives did a quick shift of watering before returning to the family
compound to cook dinner.

GENDER, ENVIRONMENT, AND DEVELOPMENT: THE WID YEARS IN RETROSPECT

In the early phase of The Gambia's market garden boom, developers saw
the gardens as a means of simultaneously addressing acute nutritional
and financial needs experienced by The Gambia's rural population. Horticultural support programs were thus construed as a direct response to
the economic crisis produced by a succession of droughts. In human ecological terms, rural families in the North Bank garden districts gradually
shifted emphasis away from rainfed crops and toward a greater dependence on irrigation from groundwater sources. This shift hinged almost

solely on the deployment of female labor. At the same time, in *political* ecological terms, women became primary providers in many households. Thus rural families thrived or suffered on the basis of gardeners' ability to generate incomes from production in lowland ecologies; they were fed and clothed on the basis of gardeners' success in surmounting seasonal barriers to expanded production. In effect, the women-environment relationship was restructured as women reclaimed lowlands for intensified dry season vegetable production.

Beginning in the mid-1980s, the heavy emphasis on "women in development" as a primary ideology motivating particular forms of development intervention began to wane. A new development paradigm centered on the environment came to the fore, invoking environmental problems of global proportions and promulgating a host of new environmental interventions. Just as with the economic diversification strategies emerging from the droughts of the 1970s and early 1980s, women had a special place in this framework. As noted in Chapter 1, these women were seen as sources of "strength and resourcefulness;" they were "promoted as 'privileged environmental managers' . . . possessing specific skills and knowledge in environmental care" (Braidotti et al. 1994, 88). They were also, as we shall see, a key source of labor in efforts to reorient rural production toward conservation and environmental restoration programs.

I turn now to explore how ideas about gender and the environment employed by developers inflected rural land use systems in the garden districts along The Gambia's North Bank. I begin with a narrative of the localized land tenure dynamics that spun out of the WID years. I show how first gardeners, and then landholders, gained the upper hand in a struggle to control garden plots, a story in which the practice of tree planting in gardens played a key role. I then consider these same dynamics from a much broader analytical perspective in order to explain how and why developers became involved in undermining women's gardens under the guise of agroforestry projects. In brief, the second half of the book is designed to tell the story of how women's gardens became men's orchards, all in the name of environmental stabilization.

Branching into Old Territory

*The Gender Politics
of Mandinka Garden / Orchards*

THE NATURE OF USUFRUCT:
RANSOMING GARDEN LAND

The key to the various strategies rural Mandinka women developed for horticultural intensification lay in the ability of women to acquire additional land resources to grow their crops. Until the garden boom, the dominant land tenure system in and around Kerewan was centered on cultivation of rice, groundnuts, and the coarse grains (millet, sorghum, and maize). Upland areas cultivated by men in a groundnut / coarse grain rotation were known locally as *boraa banko,* or "land of the beard." So-named, according to one informant, because it is "something a woman will never have," *boraa banko* land was inherited along patrilineal lines. By contrast, most of the arable swampland lying along the main river and its tributaries was controlled by women rice growers.[1] As such, it fell under the classification of *kono banko,* "land of the [pregnant] belly," and was transferred directly from mother to daughter (or daughter-in-law).

Typically, a narrow band of low-lying land formed a boundary between these two zones. Technically *boraa banko* in most cases, the soils were at once too heavy for groundnut production and too dry to support a rice crop, especially in drought years. It was in this zone that women requested parcels of land for gardening purposes from the small group of male elders who controlled the land on behalf of their respective lineages.[2] Virtually all of the eleven communal garden sites analyzed in this chapter operated on the basis of usufruct land grants issued by senior male members of founding lineages in Kerewan. Although the gardens, which ranged in size from a fraction of a hectare to nearly 5 ha were often sited on *boraa banko* land named after male landholders, the degree of authority these men retained with respect to land use practices in the gardens eroded markedly following the onset of the boom. Despite nominal male oversight, the planning and supervision of day-to-day operations were gradually assumed by leaders of the women's groups who worked the land. It was they who organized regular maintenance functions such as fence repairs and seasonal land clearing operations and they who levied fines against group members for offenses such as failing to prevent livestock from entering the gardens. The full extent of the control women asserted over their gardens on both individual and collective bases must, however, be assessed with reference to three additional aspects of garden tenure: rights of plot transmission, rights of development, and rights to tree planting.

Whereas garden perimeters were managed communally as described above, most women worked their own land allotments (avg. 300 m²) individually, or with a small group of female relatives. Two key questions, then, in understanding the tenure dynamics of the garden districts are how each individual woman came by her plot rights originally, and whether she was free to transfer those rights to her daughter or other female relatives. In 1991, I compiled the history of tenure change in 274 plots located in 11 communal garden perimeters in Kerewan, including the origins of usufruct claims (table 13) and all transfers of cultivation rights to subsequent users. I also interviewed the male landholder on each site to determine what conditions, if any, were placed on access to plots. The results of these surveys indicate that the pattern of plot acquisition changed considerably over the course of the garden boom and that many of the plots were acquired in direct contravention of conditions stipulated by landholders. Findings show that small groups of gardeners were granted use rights to plots by virtue of preexisting claims to rice land,[3] as gifts from landholders who were related by birth or marriage, or as temporary loans from other gardeners. The histories of the vast majority of plots, however, mark more substantive shifts in the nature of landholding and usufruct claims in the garden districts. Three types of acquisition—via claim payment, unauthorized gifts from female relatives, or exploitation of openings created when existing gardens expanded—bear closer scrutiny.

TABLE 13   SOURCE OF WOMEN GARDENERS' LAND
USE RIGHTS, KEREWAN, 1991

Category	No. of Plots	%
Preexisting claim to rice land	5	2
Gift from landholding male relative	6	2
Temporary loan from other gardener	17	6
Claim payments paid to landholders	128	47
Unauthorized gifts from female relatives	64	23
Plots created through site expansion	54	20
Total	274	100

SOURCE: R. Schroeder, survey data.
NOTE: *n* = 274.

The most common means of access to land for women gardeners
in the Kerewan area was via a one-time cash payment to landholders.[4]
The local term, *kumakaalu* (pl.; sing., *kumakaa),* once meant the ran-
som payments made to free family members stolen or captured into
slavery or a bail payment to free someone from jail.[5] *Kumakaalu* were
commonly assessed by landholders early in the garden boom, when
many of the area's horticultural perimeters were founded. The nature
of the tenure hold granted under *kumakaaroo* (the practice of granting
kumakaalu) was disputed by my informants. Most landholders saw the
transfer of use rights to women under *kumakaaroo* as a temporary
arrangement. As one landholder put it: "When he asked whether I sold
the [garden] land to [the women] . . . I replied that I *lent* the land to the
people to work." Landholders maintained that the money collected was
used to help pay for fence repairs or to defray other expenses incidental
to the garden's upkeep. By contrast, gardeners claimed that funds paid
to join garden groups were routinely diverted to the landholder's personal use,[6] and that *kumakaalu* thus constituted lease payments. In the
words of one woman, "As for garden land, we *hire* that from the landholder." This interpretive dispute notwithstanding, it is clear that land
transactions in the early stage of the garden boom were widely mone-
tized and that *kumakaa* payments constituted a form of disguised rent
appropriated by landholders as land was "ransomed" for gardening
purposes.

The *kumakaaroo* claim system included the proviso that each time a
plot was vacated due to the death, retirement, or relocation (due to mar-

ital status change) of its original occupant, it had to be returned to the landholder before reallocation. This condition was set in order to give the landholder a chance to exact a new *kumakaa* payment from any prospective gardener before granting her leave to put the plot into production. In addition to the financial windfall, this resumption of plot control was meant to symbolically underscore the landholder's residual land claims. This stipulation notwithstanding, the detailed plot histories gathered in Kerewan in 1991 reveal that, in practice, women often flouted this convention.

In the survey, each plot holder was asked to indicate whether she was an "original" plot holder or whether her plot had changed hands since the site's enclosure. She was then asked whether she had been required to pay a *kumakaa* before beginning cultivation. Results were sorted by location and compared with the tenure conditions landholders claimed to be enforcing in each site. Of the 274 plots in the survey, 160 remained in the hands of their original claimants. Of the 114 plots changing hands since the onset of the boom, only 33 reverted to the control of the landholder before being parceled out a second time, and only 27 of these were ultimately reallocated on the basis of a second *kumakaa* payment (6 were awarded by landholders to family members or acquaintances free of charge). Thus, the vast majority of the plots changing hands (81 of 114, or 71%) were passed directly from one woman to another, *without* any form of direct compensation for the exchange of use rights.

This group of 81 plots needs to be differentiated further in order to get a clearer picture of the property dynamics in play. First, 17 plots were allocated on a temporary basis. Women gardeners who were pregnant, caring for a newborn, or simply too ill to work occasionally loaned their plots to relatives or close friends. In most cases these were seasonal loans, but they could also be longer term, in which case the likelihood that they would eventually result in full transfers was high. A second group of 25 plots changed hands in the one large garden site in my study where such transactions were not proscribed. This garden differed from other sites in Kerewan in that the landholders ceded all forms of control to the women's group leaders shortly after it was founded. Indeed, group leaders proudly boasted that their garden was the only true "women's" garden in town. Thus the whole garden—a 3.5 ha site, or more than 10 percent of the total enclosed land area under garden production in Kerewan—was removed from male control. Finally, and perhaps most telling, is the fact that the 39 remaining plots in this group of 81, or

roughly one-third of all transfers of plots in the research sample between
1973 and 1991 (39 of 114), were permanent transfers which took place
surreptitiously (i.e., without being explicitly sanctioned by the landholder in question).

That surreptitious transfers of plot rights could take place in such a
small community (est. pop. 2,500) might seem unlikely, but for two factors. First, women's groups had come to dominate gardens so thoroughly in the 1980s that the space enclosed within the fences became
a kind of terra incognita for men. Even putative landholders were rarely
seen within the fence perimeters because of the discomfort they felt in
the midst of such a clearly defined women's space. Thus, the determination of who was actually cultivating a given plot was sometimes difficult
to make. The situation was made more complex by the fact that as many
as a third of the work units in the sample revolved around mother–
daughter tandems (table 11; fig. 6). Younger women often assisted their
mothers with their gardens and were thus in a position to gradually assume practical control over plots over the course of several years. Use
rights were accordingly established on a de facto basis, *kumakaa* "ran-
som" payments were not required, and challenges to the younger
women's claims were highly unlikely. In sum, land that was once clearly
male-controlled (*boraa banko* land) was taken over by female vegetable
growers and was being managed as though it were part of a *kono banko*
(matrilineal) legacy, a step many women felt was amply justified:

A woman has the right to give her garden land to her daughter. . . .
Suppose I spent a lot on my garden, and after growing old, I don't
have the energy to continue the work. Then in all fairness my daughter
should continue to benefit from what I spent. That is no problem, since
I toiled for it, and I also spent money on it, my daughter can have it.
That is no problem.   *Kerewan gardener*

The final means of establishing access to garden land involved the openings created when an *existing* garden site was expanded. For several
years in the 1980s, garden projects dominated the activities of NGOs
and voluntary agencies working in rural Gambia. The generally industrious attitude of women's groups, coupled with the prospects of simultaneously addressing goals of income generation and nutritional en-
hancement, made gardens attractive investment targets. Not only did

Figure 6.   Mother–daughter work tandems. The social organization of
market-garden production into mother–daughter work units has facilitated
intergenerational land transfers. These arrangements have allowed gardeners
to avoid lease payments, thereby eroding male landholding privilege.

the gardens quickly yield profits, they also helped NGOs address deep-seated social inequities. They were therefore prime targets for funds generated under WID initiatives. The intense focus on horticulture during this period meant that women's groups were often in a position to leverage successive grants to support their horticultural efforts. Whenever a new grant resulted in expanding fence perimeters, women already active in the gardens awarded themselves "expansion plots" *(lafaa rangolu),*[7] extending their use rights *without* paying additional *kumakaalu.*[8] Such expansion came at the expense of *boraa banko* land claims and accounted for 20 percent (54 of 274) of all plots in the sample.

The lack of opposition by landholders to such arrangements marked the fact that the community's "moral economy" (Scott 1976)—the fluctuating sentiments of community members regarding notions of communal benefit and well-being—had shifted in favor of gardening. After initial resistance, most men in the garden districts had "seen the benefit" *(nafaa)* of gardens in the form of cash gifts and other financial support from their wives and became firm supporters of the garden enclaves. Moreover, the offer of material assistance by NGOs and expatriate volunteers helped shore up the gardeners' claims. In theory, landholders might have refused individual requests from women to expand their plots, but the stakes for the community as a whole were high. Refusal to extend the women's usufruct rights would have meant denying gardeners access to additional donor assistance. Thus, for the most part, the few relatively senior men who controlled low-lying lands were initially hesitant to block expansion, despite the loss of power and prestige accompanying the loss of territorial control.

To sum up, the practice of *kumakaaroo,* which was nearly universal in Kerewan in the early stages of the garden boom, was only upheld in 24 percent (27 of 114) of the land transfers that occurred during the eighteen years covered by my survey. More than 70 percent (81 of 114) of the cases of plot transmissions had been negotiated directly between women, often without the knowledge of landholders. At the same time, rights to a fifth of *all* plots under cultivation were effectively leveraged through the intercession of NGOs promoting the expansion of existing garden projects. On balance, this was a striking loss of privilege for men who were accustomed to controlling the distribution of benefits generated by development interventions and who now found themselves virtually frozen out of development altogether. Some sort of move by landholders to try and reclaim their former position was virtually inevitable.

THE GENDER POLITICS OF SANYANG GARDEN

The critical break point surrounding the expansion of usufruct rights came in 1984, when a local landholder challenged the efforts of Gambian extension agents and an expatriate volunteer to secure funds to redevelop a garden site on *boraa banko* land he controlled. The 4.3 ha garden along Jowara Creek *(Bolong)* was one of the oldest and largest communal sites in Kerewan. The landholder in question, whom I will call here Al-Haji Abdou Sanyang,[9] was actually only one of three or four men who held rights there. His was the largest block of land, however; the garden was consequently named after him, and he was clearly the dominant figure in negotiations over the site for most of its history.

Development of the Sanyang garden began modestly enough. Government extension agents contacted the landholder and asked him to select thirty women to participate in a new program to promote production of bulb onions. At the time, in the mid-1970s, women on the site worked in small, individual plots, each of which was separately fenced with poles, thorn bushes, and woven grasses; irrigation wells were exclusively hand-dug and unlined; and most sales of produce were carried out on a strictly local basis due to poor roads and a lack of market outlets. Hybrid vegetable varieties were virtually unknown. In fact, according to Al-Haji Sanyang, at the time of the garden's foundation, no local growers even knew that onions could be grown directly from seed. Moreover, when a few cabbage seedlings were distributed by extension personnel for transplanting, growers used string to tie them into tight little bundles, thinking that was the only way the plants could ever form heads like those they had seen in the market.

The fledgling nature of these early attempts to establish commercial horticulture did not escape the attention of men in the village who were roundly disparaging in their initial assessments of the scheme. No one, the men concluded, could seriously expect to earn significant income from gardens, and the replacement of the patchwork of individual plots that predated project activity with a communal fence was scorned as "folly," a waste of time and money. The women were "gone to their second husbands," and as far as the men in Kerewan were concerned, these new husbands could have them. Despite this ridicule, women took to onion gardening quickly and their participation grew steadily. Indeed, each woman went so far as to pay a small fee to Al-Haji Abdou to establish claim to her plot and contributed a quota of fence posts to assist with construction of a communal enclosure.[10]

The quick growth of Sanyang garden was due in part to the reversal of fortunes in the male and female cash-crop sectors as the economic squeeze described in chapter 2 began to take hold and households became increasingly dependent on female income. The domestic budget situation of the family of Momodou Keita illustrates just how painful this transition was in many households. The residents of the Keita family compound in 1991 included Keita himself, a man in his late sixties who was the compound head, his three wives, Bintou, Umi, and Mariama, all well into their mid-fifties, a son, Dembo, his wife, Lisanding, and several small children. A second son held a civil service job in the capital, and two other grown daughters were married in nearby villages. The older women and the old man's daughter-in-law all worked in Al-Haji Sanyang's garden. However, the younger son's wife had borne three children in as many years and had not yet accumulated access to sufficient garden land to make a significant contribution toward meeting the family's cash needs. Instead, she was assigned responsibility for many of the heavier domestic chores such as providing water for the entire compound, pounding grain, and doing laundry.

Although the responsibility for provisioning the household was still nominally in Momodou's hands, in practical terms, the farming load was now carried by the son who resided at home, and the financial burden for providing supplementary grain rested on the shoulders of the elder son working in the capital. In other words, the household was in the midst of what might be seen as a normal demographic transition, and the economic contribution of the elder Keita had become increasingly marginal. This transition intersected with stagnation in the groundnut economy and a period of highly volatile rice prices. Consequently, the sons were occasionally unable to meet all of the family's staple grain needs, and Bintou, Umi, and Mariama had to step in and assist with the burden of household reproduction. From Momodou's perspective, this was an embarrassing turn of events. When I asked him how he managed his family's food needs, he produced the bag of rice and the cup he used to dole out portions to his wives for cooking each day. When pressed to describe what happened when the rice and millet grown by his son and wives ran out, he acknowledged that his wives would all buy supplemental grain with their garden money. He insisted though that he, himself, never ate this rice. Instead, he explained, when the family's home-grown supply of staples was depleted, he would send a message requesting a new allotment from his son in Banjul. While waiting for it to arrive, he would feign dysentery and refused all meals.

When Keita's three wives were interviewed, they provided a some-what different gloss on the Keita family financial situation, however. In separate accounts, they indicated that their role in meeting household grain needs was not confined to occasional and extraordinary interventions but had become quite routine. Each of the wives purchased rice to supply the family on a regular basis for much of the year, despite the fact that they had passed their prime productive years as vegetable growers. They all spoke somewhat disparagingly of their husband's inability to meet his obligations to his family but saw little alternative given the prevailing set of economic conditions.

With such household dynamics as a backdrop, the gardeners in Al Haji Sanyang's garden continued intensifying production, and it was not long before their efforts came to the attention of outside funding agencies. In 1978, the first of two major consignments of material support arrived, including tools, chicken wire for fencing, and cement for lining wells. These materials were used to expand the perimeter to allow more women to join the garden and replace the existing fence with a sturdier design that was more effective in preventing livestock incursions. At the same time, the material grants signaled a significant new point of departure in relations between growers and the landholder. Allegations made several years later in a deposition submitted by one of the growers' group leaders to the executive secretary of the National Women's Bureau specified the source of the growing problem:

> We got several helps from donor agencies and the agriculture department, all come in the form of materials and [Al-Haji Sanyang] kept the all and brought very little proportion of these things to the garden, we didn't querry then as we were all eager to work on the garden. He kept the best part of our supplies to himself [so] that sometime we had to contribute to buy chicken wire . . . we provide our own poles for the fence and buy nails from our subscriptions. (Project files, 1984)[11]

The allegation contained in this account was that the landholder took advantage of the sudden availability of developmental largesse to siphon off a rent above and beyond the claim payments he had already received. This tactic left growers in a dilemma. They did not want to call attention to the landholder's actions lest their challenge result in the loss of relatively short-lived and shaky land use rights. At the same time, the alleged diversion of funds and materials intended for the group's use could not continue indefinitely if the group was to successfully intensify production.

THE PRO-GARDEN ALLIANCE

By the early 1980s, the household budgetary burden carried by Kerewan's gardeners had only increased as the groundnut economy continued its decline and structural adjustment reforms imposed by the International Monetary Fund and the World Bank raised the cost of food (McPherson and Radelet 1995). These developments created pressure on growers to expand the area under garden production. Accordingly, when an expatriate volunteer was posted in the village in 1983, local growers' organizations seized the opportunity and lobbied the volunteer for additional material support. Representatives of the two largest grower groups, extension agents from the civil service, and the volunteer met several times over a period of months to draw up plans for the construction of some seventeen wells and expansion of existing perimeters sufficient to accommodate 479 women, or virtually the entire adult female population of the town.[12] Funding was to be provided by a foreign donor, and the project itself was to be jointly administered by the volunteer and the Department of Community Development (DCD). The Ministry of Agriculture would be responsible for surveying and allocating individual land parcels, and project materials would be temporarily kept in the Ministry's storage facilities until they were required at the sites themselves.

This last provision proved to be a sticking point with Al-Haji Sanyang, however. Although he had been a strong backer of the project in its early stages, he balked at the storage arrangement and objected to the women signing a project agreement on their own behalf with the donor agency. He had acted for a number of years as advisor to the group and, therefore, felt he should have been allowed to sign the contract as the project beneficiary. The extension agent assigned to the case described the attitude of Sanyang in explicit gender terms: "As men among women, they feel that they can dictate at their own pace what the women should be doing. Totally they are against women signing for the materials that are coming" (Project files, 1984). In the end, a compromise was struck, according to which both the landholder and the women's group leaders signed the document, but this did little to alleviate the growing tension.

The Women's Bureau deposition picks up the story from there:

> Our difference with [Al-Haji Sanyang] came when the materials arrived, because he was thinking these materials will also be taken to him for storage but instead our donors . . . said they should be kept in one safe place and we shall be getting our supplies as we work [i.e., materials were to be doled out

on a regular basis as needed]. The volunteer has chosen the store of the agric department at Kerewan.

> [Discovering this, Sanyang] turned otherwise saying that he is going to root us all out because the land is his and the first things the donor agencies have given to us [prior material grants made to the group] were his[,] for those were given to him [for safekeeping] and not to us, forgetting about the D5.00 each [the claim payments used to gain access to land; about $1.75] we have given to him and that those things were given to the women's garden and not to him. (Project files, 1984)

The issue, once again, was the control of a potential source of disguised rent, but where the landholder had allegedly been successful in skimming materials in the past, the gardeners blocked him in this case by convincing civil servants to store supplies in government facilities. Frustrated in his attempts to control construction materials outright, Sanyang then focused his attention on the proposal for awarding well-digging contracts, questioning the decision to hire trained technicians from within the government rather than allowing him to handpick subcontractors from the village. His close scrutiny of this and other decisions regarding project cash flow eventually led to an altercation at a meeting of the garden group leadership. Insults were traded between Sanyang and the project extension agent, mutual accusations regarding financial motives were exchanged, and the two nearly came to blows.

The key point of interest here is the confrontation between a powerful member of the community's landholding elite and a new set of interests—growers' husbands, state functionaries, and volunteer representatives of foreign development capital—allied in support of the women's group. Sanyang was a locally prominent member of one of The Gambia's major political parties (since abolished by the 1994 coup). Several party stalwarts rallied to his defense in the Kerewan dispute. In addition, he was supported by members of his own landholding lineage, including some of the female gardeners in his perimeter. Meanwhile, growers' husbands supported the majority of the women, both as a way of expressing their political opposition to Sanyang and as a means of protecting the basis of their families' livelihoods. Thus the controversy was not drawn exclusively along gender lines. It is striking to note how systematically the landholder probed the new pro-garden alliance for weak spots by first making the contract an issue, then challenging arrangements to store materials and award subcontracts, and finally threatening to throw women off his land altogether. The women gardeners, for their part, proved themselves adroit in the use of the institutional framework afforded by the civil service and the funding agencies to protect project assets from being seized a second time. Under these rapidly changing circumstances, it was clear that the basic parameters of the underlying landholding rights controlled by Sanyang and the usufruct rights held by the women were put to severe tests.

STATE INTERVENTION

The precise sequence of events that transpired over the next several weeks is difficult to reconstruct from available documents and oral histories. The divisional commissioner for the North Bank Division, who was resident in Kerewan at the time, became involved at several key junctures as the political factions supporting Sanyang and the women's group, respectively, took shape. At one point, the commissioner appeared to side with the women's group, going so far as to decree that a formal lease agreement should be signed ceding all land rights to the growers. Later he changed tack, ordering a halt to all construction and recommending that the women seek an alternative location for their garden activities. While this latter suggestion was a politically expedient solution to the conflict, growers were reluctant to abandon plots and wells they had painstakingly maintained and improved over several years. Moreover, Sanyang garden was situated in a prime location. It was relatively close to the village proper, had easy access to groundwater, and was situated along one of the town's major roads, a boon to efficient transportation of produce.

The response of the growers to the commissioner's directive was twofold. On the local front, several of the more militant growers designated a communal workday in order to proceed with plans to tear down the old garden fence and relocate it, vowing that they would only be removed from the garden by force. This tactic drew the attention of the police, who arrived at the site on the appointed day and confronted the extension agent working with the women, threatening to arrest him if he "moved a single stick of fence post." The agent pointed out that it was the women, and not he, who were uprooting the fence, and the police responded by taking three of the leaders of the garden group into custody. This prompted an immediate mass demonstration on the part of well over a hundred women, many of whom marched single file over a kilometer to the police station to indicate their solidarity with their leadership. The more vocal and militant among this group demanded that they, too, be arrested since they were equally culpable of the crime of refencing their garden. As word got out that garden leaders had been arrested, hundreds of other women from other sites surrounding Kerewan reportedly streamed into the police compound to observe the confrontation. Faced with a situation growing quickly out of hand, the police maintained that they had no official position regarding the underlying land tenure dispute, but that they had a right to intervene to protect the peace. Despite last minute protests that plots had already been cleared for transplanting and fervent denials by women of violent intent, the police persisted, apparently with the commissioner's backing, in issuing an injunction to ban gardening altogether on the site until further notice.

Failing in their attempts to force a solution to the dilemma on the ground in Kerewan, the growers' group then carried their case to national authorities. A deposition (cited above) was sent to the Women's Bureau, a branch of the president's office set up to address women's concerns and direct WID funding, and to the director of the organization supporting the expatriate volunteer. The volunteer's supervisor eventually intervened on the grower group's behalf to pressure authorities to allow the project to proceed. After a cooling off period of several months' duration, all the principals to the dispute were called together by the district commissioner to air their grievances in court, and a final ruling on the case was issued. The judgment on the case stipulated at the outset that the leaders of the growers' group should apologize for "publicly humiliating" Al-Haji Sanyang, and it bid them to abide by his authority over land use practices within the fence perimeter in the future. However, while this pronouncement saved face for the landholder, it did little to alter the proposed plans for developing the garden site. It stopped short, for example, of granting the landholder access to project materials, and it upheld the right of project administrators to issue the well-digging subcontracts to artisans of their own choosing. There were also no restrictions placed on the expansion of the garden perimeter, and, despite Al-Haji Sanyang's threats to the contrary, garden group members with use rights to previously allocated plots were allowed to continue cultivating their crops unimpeded.

The only exception to the rather sweeping victory by the vegetable growers concerned the issue of tree planting in the garden. Al-Haji Sanyang had complained that vegetable growers had planted dozens of fruit trees within the perimeter without his clear authorization, and he insisted that they be removed, a position that the court supported.[13] From the vegetable growers' standpoint, the practical effect of this last finding was immediate. Within a day or two after the commissioner's decision was announced, Sanyang visited the garden and ordered several dozen trees removed. In a few cases, women were granted reprieves on the grounds that they or their husbands were members of Sanyang's lineage and could, therefore, assert a weak claim to use lineage land for tree-planting purposes (Osborn 1989; see discussion below). In other cases, women begged the landholder simply to take over the trees himself rather than destroy a mature source of fresh fruit produce. A third group of growers seeking to preserve some benefit from the labor invested in their tree seedlings tried transplanting their trees out of the garden site, with little success.[14] Finally, a fourth group of growers opted for a more militant stance vis-à-vis both the landholder and the court ruling by flatly refusing to remove their trees. One grower explained, while showing me a stump of a mango tree that was cut down after the decision, that she simply told the landholder: "If you want that tree removed, there's the cutlass; do it yourself. Because there's no way you will ever make me do it for you."

Sanyang's display of the landholding privileges embedded in Mandinka customary law and reinforced by court decision did not end with the tree removals, however. Instead, in an action that foreshadowed much of what was to come in Kerewan's gardens, he immediately planted several dozen trees of his own within the perimeter. Enlisting the technical assistance of a government forester trained in agroforestry techniques, he carefully located citrus and mango trees within beds already allocated to vegetable growers. In doing so, he asserted that the initial grant of usufruct to gardeners had not erased his own use rights. Moreover, in making this locational decision, he effectively advanced a claim on female labor. His expectation was that, by planting the trees directly on top of vegetable beds, they would survive off the hand-drawn water applied to the plots by growers for irrigation purposes.

The significance of these events for my present purposes lies in what they tell us about the changing character of landholding privilege and usufruct rights during the garden boom. Over the course of this highly controversial year, the landholder, a senior member of one of the town's founding lineages, was forced to claim and defend several specific rights to land he purportedly controlled under *boraa banko* conventions. By constantly shifting tactics, he was able to probe the pro-garden alliance, testing the resolve of the various actors to see whether he might forge alternative ties that would allow him to reassert his control over the newly valuable land resources. Several of the claims he raised (e.g., that he should have decision-making authority over the location of wells and fences and the allocation and withdrawal of plot use rights) were consistent with *boraa banko* rights practiced prior to the garden boom (cf. Mackenzie 1994). That he was forced to defend them at all is indicative of the extent to which his landholding status had changed since the early stage of the garden boom. Loss of practical control over production decisions was not the only issue raised in the landholder's complaint, however. Other claims, such as the right to sign quasi-legal land use agreements with NGOs and state agencies on behalf of the gardening group, the right to store construction materials intended for use in fence repairs and well digging in his own family compound, and the ability to personally award well-digging subcontracts, reflected fears that his ability to exact real and disguised rent payments from the gardens would be greatly diminished under the new production regime. If he did not resist the gardeners' efforts to establish direct ties with the NGO funding community, his own ability to convert lineage-based landholding rights—best understood perhaps as a kind of stewardship responsibility held on behalf of his kin—into the equivalent of private property rights would be threatened. His expectation that he should be free to control development largesse, extract rent from developers and garden groups, and channel benefits to family members and friends was born of his experience of several decades of development interventions that had operated in precisely that fashion (Carney and Watts 1991). When extension agents for the Gambian government and the expatriate volunteer coordinating donor support for the project refused to release funds and construction materials to him directly and proscribed many of the rent-taking mechanisms he had previously employed, the landholder balked, threatening violence and vowing to evict the gardeners altogether.

THE TREE TENURE QUESTION

The controversy in Sanyang garden in 1984 marked a watershed in the political ecology of The Gambia's North Bank Division. The involvement of hundreds of women in the demonstration at the Kerewan police station and the traumatic nature of the public court case led to an unprecedented mobilization of social, economic, and political interests in the village. As the affair became more heated, every step taken by Sanyang to challenge the pro-garden alliance, and every counter-tactic used by women to legitimize their use rights, was carefully scrutinized and debated throughout the village and beyond. This led other landholders to reappraise their own plans for management of low-lying land resources.

Most critically, the case of Sanyang garden raised the complex issue of tree tenure. In chapter 2, I noted that the Mandinka land tenure system on the North Bank maintains a basic distinction between upland areas and swamp lands. Since the virtual demise of the female-grown *digitaria exilis* (Mandinka: *findo*) grain crop in the 1940s (Gamble 1955), uplands in the Kerewan area have been a fairly exclusive male production domain devoted to groundnut and coarse grain production; swamps conversely have been used for rice production by women. Virtually all of the communal garden sites considered in this research were founded on patrilineal lands. Each site originated with a usufructary land grant from one or more of the relatively senior men in the area's founding lineages, who subsequently maintained varying degrees of authority with respect to decisions governing land use practices within the designated parcel. This power could potentially include the determination of which crops were to be grown, when, and by whom, and whether specific forms of land development such as well construction, fencing, and tree planting could take place.

Tree-planting rights are widely acknowledged in the tenure literature to lie *outside* the bounds of "secondary" usufruct rights such as those pertaining in Gambian garden perimeters (Fortmann and Bruce 1988; Freudenberger 1994; Raintree 1987). Landholders typically refuse to grant tree-planting rights to "secondary" tenure holders because they fear the longevity of trees will negate the possibility of alternative land uses. In this regard, three general principles of tree tenure incorporated within Mandinka customary law have special relevance (Osborn 1989). First, tree-planting and land use rights are partible; trees belong to those who plant them and not necessarily to those whose lands are used for tree-planting purposes.[15] Thus, although most tree planting is done by individuals on land that they control, this does not preclude entirely the prospect of planting elsewhere.[16] Second, tenure rights frequently vary according to tree species, with ownership being most clearly articulated in the case of non-native or introduced species. And third, the rights to tree *benefits* are traditionally somewhat diffuse. As Osborn explains:

> Benefit distribution is dominated by an owner's obligations to their family and community. In this regard, the concept of property as a right not to be excluded . . . is important. . . . Although the link between tree planting and tree ownership is direct and strong, the distinction of which *individual* owns

a tree is less important than which *compound* [family-based residential unit] the tree belongs to. (Osborn 1989, 66; emphasis added)

The principle of partibility was clearly illustrated by the tree-planting practices of Kerewan women in the early 1980s. Despite limited land-holding rights, 82 percent of the women controlled trees in upland ar-eas, primarily in areas immediately within or surrounding family com-pounds in town (cf. Rocheleau 1987, 1995; Rocheleau et al. 1996). Significantly, of an average 11 trees held by women in the uplands, 7.4 were inherited and 3.2 were planted by the informant herself. This sug-gests that recognition of tree-holding rights for women was both long-standing and continuing. A critical question is whether the relatively weak usufructary claims most women held to *garden* lands embraced discretionary control over tree planting on those locations. Survey re-sults indicate that women asserted such rights unilaterally: 83 percent of Kerewan's gardeners had planted trees in their gardens as part of complex agroforestry practices geared toward maximizing their returns from small plots and extending the market season for fresh garden pro-duce. While this figure obscures the important species differences dis-cussed below, it nonetheless represents once again a substantial erosion of patriarchal land use controls. Until the 1984 court case, this unilat-eral extension of usufruct rights went largely unchallenged. As the saga of Sanyang garden demonstrated, however, when Sanyang found him-self stymied on other fronts, his tree tenure rights were a key element in his attempt to reassert landholding claims.

TREE COUNTS AND GENDERED TENURE: MEASURES OF CONTROL IN GARDEN / ORCHARDS

In order to gain a broader perspective on the scope of garden-based or-chard development along the southernmost stretch of Jowara Creek and to understand the significance of species differences for tenure claims, I conducted a detailed physical inventory of the trees planted in each of twelve garden perimeters. This exercise was intended to help develop a synoptic overview of fifteen to twenty years of tree planting in the Kere-wan vicinity and assess the physical evidence of attempts to control gar-den space through tree-planting practices.

In each of Kerewan's gardens, special consideration was given to the relationship between use rights to land and the right to plant trees. In most cases, landholders specified not only whether tree planting was al-

Figure 7.   Intercropping papayas. After a 1984 court case politicized the issue of tree tenure in the Kerewan garden district, landholders restricted tree planting in many garden sites. Papaya trees were the exception to this rule, largely because they were physically less substantial and thus posed less of a threat to other potential uses of the site.

lowed but which species were affected and to whom specific injunctions and dispensations applied. Taller, woody species such as mangoes and oranges were difficult to remove and posed shade problems to underlying crops. Thus, it was common for landholders to block women from planting these trees while allowing extensive cultivation of papayas and bananas on the same land (fig. 7). Alternatively, planting of more durable "shade species" was allowed, but the rights to do so were reserved for the landholder himself.

Table 14 summarizes the situation for twelve Kerewan gardens in 1991. For each site, the table indicates the size of the perimeter, the number of years since the garden was enclosed within a communal fence, and a measure of planting density for trees found in different size and shade categories. Counts for the shadiest species—mangoes and oranges—are listed separately. Trees grouped under the heading "Other species" include the less "permanent," but numerous and economically beneficial, papaya and banana trees, as well as a small number of hardwood forest species, which for the most part predated the gardens themselves. Actual tree counts have been converted to measures of planting density for pur-

TABLE 14   TREE-PLANTING DENSITY AND TENURE
RIGHTS IN KEREWAN GARDENS, 1991
(trees in all stages of maturity)

| | | | | Tree Species | |
| | | Area | Years | | | |
Tree-planting rights	Site	(ha)	fenced	Mango	Orange	Other
Group A: Mango and	1	4.3	16	6	12	572
orange trees banned	2	3.5	14	1	3	111
	3	0.5	10	0	4	258
	4	1.4	7	3	2	140
	5[a]	3.0	3	0	0	104
Group B: Full planting	6	3.6	15	64	20	225
rights for women						
Group C: Full Planting	7	1.3	10	33	12	32
rights for men and women	8	0.2	0	n / a	n / a	n / a
Group D: Mango and	9	2.2	4	90	14	168
orange planting restricted	10	1.1	1	28	1	136
to men	11	0.4	1	45	13	48
	12[b]	1.0	1	100	40	600

SOURCE: R. Schroeder, field notes.
[a] Tree counts taken on 0.6 ha prior to 1991 expansion.
[b] Tree counts taken on 0.1 ha prior to 1991 expansion.

poses of comparison. I have also included trees at all stages of maturity, in an effort to reflect the potential as well as the prevailing status of the shade threat posed in each setting.

The twelve sites have been numbered for reference purposes and have been grouped according to the tree tenure system prevailing in 1991 in each case. Landholders controlling gardens in Group A banned tall, shady trees altogether (Sanyang garden is Site 1). Group B includes the "women's garden" noted above, in which full and exclusive tree-planting rights were granted to women very soon after the garden was established. In Group C, both growers and landholders had full tree-planting rights. Group D landholders reserved mango and orange planting privileges to themselves.

Table 14 reveals that the bans imposed on shady species were largely effective. There were very few mangoes and oranges planted in gardens in Group A. At the same time, it was clear that, regardless of any constraints stipulated or implied in use agreements, women in the early stages of the garden boom planted trees wherever they could in order to assert and secure land claims and to enhance incomes by extending the produce marketing season (see chap. 4). This was evident in gardens in Group A, where dense stands of bananas and papayas were present (see fig. 7), and in Site 6, which was the only site with no tree tenure restrictions on women. By the mid-1980s, the economic calculus favoring tree planting by women had changed, however. Markets for fresh vegetable produce had deepened dramatically along the North Bank, and comparative returns to tree planting and vegetable growing shifted decisively in favor of vegetables.[17] While no data are available from this period, Osborn (1989) reports that produce from a mature mango tree in Upper River Division fetched only D50 in 1988; in Kerewan in 1991, a year's harvest from a single tree was sold for as little as D30 (the buyer providing harvest labor). By comparison, an 8-by-8 m plot (an area that might easily be eclipsed by a moderate-sized mango tree) brought an average gross return in 1991 of roughly D240, or $30, to Kerewan gardeners.[18]

The decisive factor in gardeners' collective change of heart regarding tree crops was the emerging shade problem illustrated in the density figures recorded in table 14. The data show a marked distinction between Site 1, Sanyang garden, where dozens of mango and orange trees were forcibly removed in 1984, and Site 6, the "women's garden," which was established at roughly the same time, and in which women enjoyed full tree-planting rights. When growers in both gardens were asked in 1991 whether they experienced shade problems, 70 percent of the women in Site 6 complained that shade had a significant impact on crop yields on their plots, as compared to only 30 percent of the growers in Sanyang garden. As one of the women's group leaders in Site 6 put it: "We are afraid of trees now. . . . You can have one [vegetables or fruit trees] or you can have the other, but you can't have both."

This comment encapsulates the dilemma that faced Kerewan gardeners. Trees were once a means for somewhat surreptitiously extending use rights to land, a shady practice in its own right given the cultural norms governing tree planting and land tenure in Mandinka society (cf. Osborn 1989). By the mid-1980s, however, the relative economic benefits of tree planting and vegetable growing shifted decisively in favor of gardens, and trees became a threat to women. Growers began cutting back or chopping down trees in order to open up the shade canopy and expose their vegetable crops to sunlight (see fig. 8 and 9).

Figure 8. Emerging shade problems. In the mid-1980s, a deepening market for vegetable crops changed the economic calculus for women gardeners. Whereas they once used trees to diversify income streams and extend and solidify land tenure claims, improved market conditions meant that trees became a threat to garden livelihoods.

Figure 9. Tree-trimming practices. The aggressive trimming of trees to reduce shade problems was anathema to many developers who worked hard to
establish orchards and woodlots in the area.

BRANCHING INTO OLD TERRITORY

While gardeners gradually abandoned their tree crops in favor of the
more lucrative vegetable crops, cutting back the shade canopy further
with each passing year, landholders saw a new opportunity developing
for themselves. Male landholders like Sanyang had initially resisted tree
planting on the grounds that it reduced their future land use options, but
the availability of a female labor force to water trees, manure plots, and
guard against livestock incursions within the fenced perimeters shifted
the landholders' economic calculus *toward* fruit growing. Table 14 marks
the clear trend, beginning in the mid-1980s (and continuing up to at
least 1995, when I last visited the area—see chap. 6), toward increasing
involvement in tree planting by men. It also documents the proliferation
of mangoes, the species least compatible with an underlying vegetable
crop. Would-be orchard owners almost invariably chose mangoes as the
primary species to be planted in their orchards, a choice motivated in
part by the fact that mangoes are an important source of nutrition when
other sources of food are scarce, especially for children. The orchard
owners were also attracted by the fact that NGOs and government agencies had imported and promoted several new mango varieties. The so-

called improved mangoes fetched relatively high prices on local markets in The Gambia and across the border in Senegal to the north. They were accordingly used by developers as a means of enticing male landholders to undertake agroforestry projects as part of a broad environmental stabilization campaign.

While landholders saw the mango and citrus orchards as a source of cash income and developers embraced them as a new form of environmentally sustainable land use, the tightening of tenure restrictions and promotion of mono-crop mango production had direct and dire consequences for vegetable growers. Prior to the introduction of monocropping practices, when gardeners controlled decisions over the selection of species, the location of trees, and rights of trimming or removal, they were able to carefully manage interspecific competition between vegetables and tree crops so as to minimize the negative effects of companion planting. As soon as landholders reclaimed the initiative and developers became involved in lowland conversions, however, these prerogatives were lost, and the requirements of the vegetable crop became a secondary consideration.

Thus, a battle of sorts ensued over tree planting in gardens. Waged on a site-by-site basis, this struggle tilted in favor of gardeners on one site and leaned in the direction of landholders on the next. In 1983, Site 7 was founded immediately adjacent to Site 6, where women now experience the greatest shade problems. Given the land pressure at the time, many women from the older site took second plots in the new site. Under a somewhat novel arrangement between vegetable growers, the landholder, Forestry Department officials, and a donor agency, the garden was converted into a garden / orchard with a dense stand of trees laid out in a grid pattern over the entire area. The understanding was that ownership of the trees would be divided between the landholder and gardeners on an alternating basis; every other tree, in effect, belonged to the landholder. This was, in other words, a project that sought to "integrate" the goals of environmental restoration and economic development.

Within five or six years, shade problems began to appear on the site. Gardeners had already determined that vegetables brought them a greater return than any harvest they could expect from their trees. Consequently, many of the maturing trees on Site 7 were either drastically trimmed or simply removed, including, apparently, many of the trees belonging to the landholder. In response, the landholder banned tree trimming in his garden, only to find his young trees still being destroyed as women burned crop residues to clear plots for each new planting season

Figure 10.   Fire hazard. Orchard owners objected to the clearing of garden plots with fire because of the hazard this practice posed to young tree seedlings.

(fig. 10). While some of this destruction was doubtless accidental, the landholder claimed that growers also deliberately hung dry grass in tree branches so that fires set to clear plots would fatally damage trees. Tree density figures in table 14 reflect the fact that fully half of the original tree stand in Site 7 no longer exists, so it is clear that vegetable growers largely had their way on the site.

On other sites, however, landholders were much more vigilant. Roughly a dozen garden / orchards were established in Kerewan between 1987 and 1995. Discussions with the landholders opening these perimeters revealed that they had adopted much stricter controls over land use than those in place in many of the community's older gardens. *Kumakaa* payments, for example, were eschewed altogether. In most cases, access to land was only granted under terms that required women to guarantee they would: (1) water the tree crop as long as they stayed in the perimeter; and (2) leave their plots as soon as the trees reached maturity. With a wary eye trained on the prospect of women mounting competing claims to land or trees on moral economic grounds, the male landholders either provided fences and wells on their own account or built them with the assistance of donors interested in promoting agroforestry, many of whom had sponsored garden projects on the same sites several years earlier. In one garden destined for conversion into an orchard, a contract was signed between the landholder, the donor agency, and a garden group stipulating a five-year limit to the women's vegetable-growing rights. In another, a project manager proposed a rule as a hedge against tenure erosion that would preclude anyone other than project participants and "one small daughter per grower" (!) from working the plot. In a third site, garden gates were padlocked in recognition of full conversion to orchard production. Moreover, in 1991, the Kerewan town chief, fresh from an environmental sensitization training organized by an NGO, issued a proclamation that any gardeners responsible for starting a bush fire that destroyed trees would be fined the equivalent of $200. At least one woman was evicted from her plot when a fire she started accidentally destroyed just one of the landholder's trees.

CONCLUSION

The notion that rural African peasant women are helpless victims in the face of environmental degradation does not stack up well against the evidence drawn from the contemporary history of The Gambia's garden boom. In the previous chapter we saw how women's incomes came to outstrip their husbands' and how, after several years of struggle, they won for themselves significantly greater autonomy by virtue of the strategic deployment of their garden incomes. The financial responsibilities women undertook in order to gain these freedoms meant, however, that they were under a great deal of social and economic pressure to intensify their production efforts and boost garden output. This sometimes placed gardeners at odds with landholders on whose low-lying territory women sought to expand their plots. In this regard, the intensification of market gardening shifted the locus of the politics generated by the garden boom from a set of conflicts between women as wives and men as husbands to a series of related struggles between women as gardeners and men as landholders.[19]

The evidence in this chapter suggests that, in the latter arena, too, women were quite successful in their own terms. By means of surreptitious land transfers and tree planting, and the ability to leverage resources from NGOs and voluntary agencies, gardeners wrested partial control over low-lying lands from male landholders and gradually expanded their usufruct rights until they were in a position to manage their plots as if they were part of a matrilineal inheritance. Moreover, the Kerewan garden groups displayed an astute sense of the shifting moral and political economies taking shape both within Kerewan itself and in the development community made up of NGOs, volunteers, and sympathetic state agencies. They put this sense to good use as they successfully played the members of the pro-garden alliance off against landholders with whom they vied for control of land and developmental largesse.

The benefits of the garden boom were not restricted to gardeners, however, even in the earliest stages. Landholders, too, took great advantage of the influx of development capital to capture different forms of rent in return for their evident loss of control of low-lying territories. In the initial phase of the boom, some landholders took advantage of the access they were given to construction materials to extract "in-kind" rents from the development agencies providing support to the garden groups. As the demand for garden plots increased, the institution of *kumakaaroo* "claim payments" allowed rents to be taken from gardeners directly, as hundreds of women anted up small sums in exchange for usufruct rights. Later, when *kumakaa* payments slowed, some landholders managed to siphon rents out of subscriptions gardeners imposed upon themselves for fence and well repairs.

When women gardeners and their allies in the various development agencies succeeded in blocking these various mechanisms for profit taking, however, as was the case with the project to expand Al-Haji Sanyang's garden in Kerewan in 1984, landholders were forced to seek out other means of regaining control over resources and resuming their privileged position as arbiters of development interventions at the community level. Thus many of the newer landholders began tightening up tenure constraints on gardeners and turned to tree-planting practices as a means of tapping into the benefits generated by the boom. This time, however, the objects of desire included the well-established fence perimeters and irrigation wells, the enhanced soils, and the labor pool the gardeners themselves represented. And it was the landholders who benefited from a precipitous shift in development ideologies and practices, as the alliance that had once formed in support of market gardens regrouped around the goal of promoting agroforestry projects as a means of improving natural resource management and attaining the nebulous objectives of "sustainable development."

Contesting
Agroforestry Interventions

In retrospect, it may seem surprising that the state agencies, NGOs, voluntary organizations, and mission groups that once supported the garden boom could so blithely shift their focus from market gardens to agroforestry ventures. There were, however, at least three mitigating factors that helped steer development efforts in this new direction. First, although the garden boom produced dramatic social and economic changes in many parts of the country, this fact was not widely appreciated in The Gambia in the mid-1980s. In fact, as noted in the Preface, the garden boom was systematically discredited by several of the major developers, who saw rural women gardeners as hopelessly mired in inefficient and irrational production and marketing networks. Belittled in this way, the boom and its effects were easy to overlook. Second, in the eyes of at least some developers, the significance of the garden boom paled in comparison to the lingering effects of drought on the continent. While the late 1980s saw annual rainfall figures in The Gambia rebound somewhat from their low point in the years 1983–1985, elsewhere in Africa drought problems remained quite serious. Moreover, The Gambia itself continued to face significant shortfalls in food production, despite diverse and repeated developmental efforts to redesign the country's agrarian economy (USAID 1992). Trade figures showed The Gambia importing relatively large amounts of foreign grain, and these imports stood as testimony to the fact that the country's basic economic problems had not been addressed. As one high-level aid official responsible for his country's bilateral support to Gambian agricultural programs argued, "In the long run, gardens don't matter if we don't do something to address the more fundamental problems of sustaining successful natural resource management in the region." Finally, as the sentiments expressed by this aid official indicate, a sea change in the thinking of development officials was under way world-wide as aid agencies sought to center development efforts on the task of improving environmental management on a global scale. Culminating in the Rio Earth Summit in 1992, this multifaceted program dominated development efforts in the African region from the late 1980s on. The scope of these efforts easily surpassed the WID initiatives of the late 1970s and early 1980s. And in the case of The Gambia's garden boom, they directly superseded WID projects in many communities.

GAMBIAN AGROFORESTRY INTERVENTIONS: THE RISE OF THE TREE CROP ENTREPRENEUR

In response to the new environmental mandates, the Gambian government moved to address environmental problems along several fronts. Drought-induced salt damage in rice-growing areas, heavy erosion of beachfront properties along the coast, destruction of ground cover by seasonal forest fires, and poor sanitary conditions in the country's major cities all received attention in government environmental planning initiatives (GOTG 1992). These objectives notwithstanding, Gambian planners and their expatriate donors were particularly concerned with the declining quality of the country's forest resources.

A 1980 survey indicated that the standing stock of trees in the country's forest parks and communal lands was of increasingly poor quality (Schindele and Bojang 1995). Using the results of aerial photography and other remotely sensed imagery, analysts concluded that much of the "closed canopy" forest had converted to "open canopy" forest. Within the existing stock, there were almost no trees below forty years of age and very little natural regeneration of several of the more valuable tree species. Moreover, the incidence of dead (9%) and decayed trees (45%) was very high. While these findings were sobering enough, a second study drew the even more alarming conclusion that the deforestation rate in The Gambia was 6 percent per annum (Ridder 1991, cited in Foley 1994; these interpretations have, however, been contested; see discussion in Foley 1994). Planners argued that the apparent degradation was of critical importance for rural Gambians and had profound regional implications as well:

> At present the forests of The Gambia form the last vegetative frontier towards the desert. Because of its suitable geographic location as a long belt along the River Gambia, it would be comparatively easy to finally stop desertification at this point by safeguarding and rehabilitating the still existing forest lands. If the Gambian forests can be saved not only the livelihood of the Gambian population will be maintained, but also that of the people which are living south of The Gambia. (Schindele and Bojang 1995, 7)

As these and other research findings became available, a range of interventions were initiated in the forest sector (Schroeder). Building on legislation passed by the Gambian government in 1977 and 1978, including the Forest Act, the Wildlife Conservation Acts, and a set of Wildlife Regulations, the government entered into a series of collaborations with foreign donors, two of which bear special mention here insofar as they set important precedents for the agroforestry efforts that followed in North Bank gardens.

A particularly ambitious strategy funded under the auspices of USAID, the USAID Forestry Project (1979–1986), was designed to promote sound environmental management through the use of commercial incentives. Among other objectives, this project encouraged Gambian villagers to replace fuel gleaned from dwindling "natural" forest reserves with wood produced on small communal woodlots located on the immediate outskirts of villages. Using subsidies for wells and fences to initiate projects, promoters urged woodlot growers to manage the projects so that they would produce a sustainable income through well-timed sales of surplus wood and forest products (e.g., fodder). The promotion of community woodlots ultimately failed to produce a significant shift in village-level land use practices, however, because, analysts concurred, "the wrong tree species had been selected" (Thoma 1989, 37; see also Lawry 1988). The species favored by the project, primarily *Gmelina arborea,* neem *(Asadirachta indica),* and eucalyptus, were "wrong" because alternative fuel-wood species preferred by Gambians were still available. Moreover, the effort to build an environmental stabilization strategy around forest species ran up against what was termed a "time factor":

> Compared to most agricultural operations, tree growing is a long-term exercise. First benefits from fruit trees (Mango, Cashew, Citrus) can be expected in three to five years after planting; from fast growing exotic species

(*Gmelina*, Neem, Eucalyptus) in five to ten years; and from indigenous species in ten to fifteen years. Some projects in The Gambia . . . indicate that as long as villagers are able to obtain intermediate benefits such as vegetable[s], fruits, honey or forage they are more prepared to accept the delay in harvesting [of timber and other forest products]. (Thoma 1989, 44; cf. Lawry 1988; Mann 1990; Norton-Staal 1991)

In effect, analysts argued that the woodlot program failed because it simply did not carry the promotion of commodity production far enough. Exotic species could be, and were to some extent, commodified, but the economic returns they brought were still too slow in developing to warrant investment of time and capital by prospective woodlot managers.

Rights to project benefits were also unclear. Lawry (1988) deems such ambiguity a "classic" problem with woodlot projects (see also Fortmann and Bruce 1988; Raintree 1987): "Projects usually provide that benefits will be distributed among the villagers, but the formula for distribution is not thought through, or is left to be determined . . . by 'the village'" (Lawry 1988, 2). Such proposals ignore the fact that "'the village' is not the harmonious social formation that many project designers assume it to be. . . . To say that the project will benefit 'the village' overlooks the fact that villages are made up of individuals, families and groups with different goals and expectations, which are not necessarily equitably served by any given project" (Lawry 1988, 2). The Gambian woodlot program was also based on the erroneous assumption that villagers would willingly contribute labor despite the fact that the issue of benefits distribution was left unresolved. Lawry provides a case in point:

> There the village *alkalo* [head], a very entrepreneurial farmer and fruit grower, made available land for the woodlot, but was unable to mobilize villagers to participate in planting, cultivation, and harvesting tasks. While [he] attributed this to lack of foresight by villagers, it was clear that [he] was retaining tight control over woodlot management. . . . [V]illagers were uncertain over their rights . . . in relation to their labor contributions. (Lawry 1988, 2)

RESOLVING THE LABOR QUESTION

When the next major national forestry initiative was mounted by the European Community (EC) in 1986, program emphases shifted toward fruit production and resolution of the labor question, both in order to expedite tree planting. By this time, the garden boom was well under way, and low-lying garden sites had become attractive locational options for foresters because they offered secure fence enclosures, improved soils, and ready access to water. Crucially, they also contained a captive, if not wholly cooperative, labor force to water trees, manure plots, repair fences, guard against livestock incursions, and maintain wells (figs. 11 and 12). By the late-1980s, therefore, a firm consensus had formed among the Forestry Department, NGOs, voluntary agencies, and donors around the triple foci of (1) using the production of fruit commodities as the vehicle to promote agroforestry; (2) concentrating commodity-based tree-planting projects within gardens; and (3) managing the whole endeavor on the premise of being able to exploit a female labor reserve.

While the resulting mix of property and labor claims was not entirely without precedent in The Gambia,[1] the heavy emphasis on *female* labor to carry out tree-planting objectives was striking. A UNDP official gave voice to the ideological basis for pinning the hopes of Gambian environmental stabilization efforts on women: "Women are the sole conservators of the land . . . the willingness of women to participate in natural resource management is greater than that of men. Women are always willing to work in groups and these groups can be formed for conservation purposes" (cited in Norton-Staal 1991, 12). Her thoughts were echoed in promotional literature circulated by an NGO active in tree planting on the North Bank in 1991: "The involvement of women in the development process is vital in the Gambian context if sustainable development is the ultimate goal. . . . In the Gambia, our primary focus has been on women. . . . [I]n the implementation of an environmental programme in the country, *they could be deemed the most precious and vital local resource*" (Worldview International Foundation 1990, 6; emphasis added). Planning documents for a proposed multi-million-dollar USAID-sponsored Agricultural and Natural Resource (ANR) program stated: "The role of women in improved ANR is critical as women frequently have access to more marginal / degraded lands than men, and have exhibited a great deal of interest in conservation techniques. . . . Targeted conservation committees will continue to require a percentage of female members to assure women's participation" (USAID 1991, 28).

These references drawn from the development literature on The Gambia bear the marks of both the ecofeminist and the feminist environmentalist ideologies outlined in chapter 1. The assumption that women are always willing to work in groups formed for environmental

Figure 11. Labor demands and irrigation. Orchard owners relied heavily on female labor for a variety of unpaid services, especially the watering of tree seedlings.

Figure 12. Labor demands and fence maintenance. Landholders were
also dependent on gardeners to maintain fences and guard against livestock
incursions.

management purposes clearly naturalizes the relationship between women and their environments, and the designation of women as The Gambia's "most precious and vital local resource" makes that identification complete. The USAID statement apparently marks a more critically informed position by invoking tenure relations, but the social relations embedded in the tenure system are euphemized. It is not that women simply have access to marginal lands but that they are *limited* to such lands. As the case of Sanyang garden illustrates, the determination of tenure rights is often a very deliberate process, requiring active choices to restrict individuals from specific patterns of use.

Clearly the invocation of women's role as environmental managers, derived in part from a distorted feminist environmentalist ideology, was somewhat "opportunistic" (Rocheleau 1995, 10) in this instance. Indeed, the specific *locational* decision to target women's gardens had an immediate and instrumental political-economic rationale. One developer summarized the prevailing sentiment: "Women are reportedly *not good at* watering the trees unless they are located directly in the garden and receive water indirectly when the vegetables are watered" (quoted in Norton-Staal 1991, 6; emphasis added). To say that women were "not good at" watering trees is once again to resort to euphemism. The blunt reality is that they often simply refused to water trees located outside of their gardens, or did so only half-heartedly, when their "participation" in tree-planting ventures failed to generate direct personal benefits.[2] This pattern of resistance led forestry planners to conclude that, if women gardeners were to prove a reliable source of labor for orchard development, the task of watering landholders' tree seedlings could not be made wholly voluntary but had be imposed upon them.

A case in point was a project funded by a Norwegian agency active in youth-oriented programs and a handful of garden / orchard projects in The Gambia. In 1991, as part of a plan to help the agency's local branch meet the recurrent expenses of its regional office in Kerewan, the agency funded a 3 ha garden / orchard project. Unlike the agency's other initiatives, however, this project was conceived by and for agency personnel. One of the managers of the group, a Kerewan resident, was instrumental in helping the agency gain access to a prime section of low-lying land on the immediate outskirts of Kerewan for project implementation. The catch was that a portion of the proposed site was already under cultivation by a group of gardeners (Site 5 in table 14). Nonetheless, rather than negotiate for use of land *adjacent* to the pre-existing plot, or invite gardeners' input into the planning process, project planners

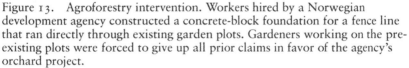

Figure 13. Agroforestry intervention. Workers hired by a Norwegian development agency constructed a concrete-block foundation for a fence line that ran directly through existing garden plots. Gardeners working on the pre-existing plots were forced to give up all prior claims in favor of the agency's orchard project.

simply swept the older perimeter off the site altogether (fig. 13). All rights to well-manured plots, wells, fences, and trees were thus summarily vacated. The land was resurveyed, and a larger group of women, including those dispossessed of their previous holdings, was invited back into the garden. Smaller plots designed according to optimum requirements for tree spacing were then allocated under strict new conditions. As one of the project managers described it: "We [Project management] will take the advice of agriculture [horticultural extension agents], not the women. . . . The garden will not belong to them, but will belong to us. . . . The [women's] plots will not be permanent; [they] will be temporary. . . . [Management will make sure that the women] take great care of trees; . . . *the survival of trees is their first goal.*" As for the potential for resistance to the agency's plans, the manager stated flatly, "We will sow the seedlings on their plots whether they like it or not."

A second case illustrating the new attitudes developing toward gardeners involved a garden / orchard site opened in 1992. The landholder in question established his orchard with the assistance of free tree seedlings and extension advice provided by the Forestry Department. Women were invited to work gardens on the site, but no *kumakaa* payments were solicited and none was paid. Moreover, no assistance was required of the women for maintenance or site protection, apart from irrigating trees as they watered their vegetable plots. After the landholder proudly provided a tour of his plot, I asked him what the women gardeners sharing his land would do when his trees matured and a shade canopy closed over the site . . . whether he would drive (*bai*) the women off his land or let them stay on. His response was most telling: "*I* won't drive them," he said. "*The trees will drive them. The trees.* If not for the trees, they could stay here."

This was a remarkable admission. The landholder's claim that a "natural" species succession was responsible for driving women off the land masked his own explicit intention of displacing the gardeners when the orchard matured. Moreover, it obscured the all-out effort by NGOs, donors, and Forestry Department extension agents to "improve" the gardens through agroforestry practices. Like the women gardeners before him, the landholder shrewdly seized an opportunity created in the context of a changing development agenda. His admission clearly revealed that tree cropping allowed landholders to develop orchards without paying full cost for labor and, at the same time, reclaim low-lying land without *directly* evicting gardeners (cf. Rocheleau and Ross 1995). Moreover, by letting the trees "drive" the women, landholders and their donor / sponsors followed the contours of a moral economy wrought by the garden boom, carefully avoiding the ill will of communities grown heavily dependent on garden income.

The single-minded focus on trees and tree crop commodities exhibited in these two cases grew directly out of the general ethos of commercial production. The tendency to disregard vegetable growers' rights and livelihoods was expressed in the tightened tenure constraints imposed by Kerewan's tree crop entrepreneurs who let trees "drive" away gardeners, the misrepresentation of women's resistance to orchards embodied in the claim that women were "no good" at watering trees outside of gardens, and the blunt assertion by the agroforestry manager that he would "sow the seedlings on their plots, whether they like it or not." In short, the new form of agroforestry introduced into Kerewan's garden district was no longer the system of shared tree benefit distribution described by Osborn (1989), but something new—a more distinctly privatized production system that pitted two ostensibly progressive development strategies focused on gender equity and environmental stabilization against each other. By promoting agroforestry, developers effectively favored the private profits of a landholding elite over other development objectives and undermined in the process a livelihood strategy that helped thousands of rural Gambian families adjust to financial hardship over the better part of two decades.

UNRESOLVED LAND USE CONFLICT

While the concerted backlash against women's gardens by male landholders described above seemed to point toward either the decline or the demise of the women's garden sector, evidence gathered during a return trip to the Kerewan garden district in 1995 showed that the outcome of conflicts over low-lying land and water resources in the North Bank garden districts was much more complicated. On closer inspection, it became clear that gardeners threatened by orchards had developed a variety of mechanisms for resisting orchard projects. Moreover, some of Kerewan's most prominent garden landholders had made surprising concessions to the garden groups. The net result was a highly varied social and political ecological landscape embracing gendered patterns of land use and control that tilted in favor of gardeners on one site and back toward landholders on the next.

By 1995, the property dynamics of Kerewan's garden / orchard agroforestry system had advanced to such a stage that one of the area landholders had reclaimed a 2.2 ha garden site in its entirety. Established in 1987 on land controlled by a landholder in neighboring Swaray Kunda, but cultivated by both Swaray Kunda and Kerewan women,[3] this site was in many ways a model agroforestry project. Most important, it boasted an energetic landholder, Lang Swaray,[4] who firmly embraced the idea of commodity-based agroforestry. Lang Swaray was a key figure in the introduction of agroforestry practices on the North Bank, in part because of his involvement with the Methodist Mission Agricultural Program (MMAP), one of the principal agencies involved in promoting agroforestry in The Gambia (Mann 1990). In 1982, the Methodist Mission established one of the country's first and largest nurseries for the mass production of tree seedlings. Based in Brikama on the South Bank, the MMAP produced several different improved mango varieties as well as shrubs, thorn bushes, and trees suitable for replacing hardwood fence posts and barbed wire as "live" fencing materials.[5] Although the MMAP nursery was a tremendous resource, the mission had difficulty servicing North Bank farmers and NGOs due to the difficulty of transporting seedlings to North Bank locations. When the MMAP subsequently sought to establish a second nursery in the Kerewan area, Lang Swaray was instrumental in helping procure land on the outskirts of Swaray Kunda for this purpose. In exchange for this assistance, the Mission offered him seedlings at a subsidized rate, which he then used to establish his orchard.

True to form, the site was set up initially as a garden, although, in keeping with the trend toward tighter tenure restrictions on garden orchards outlined in chapter 5, women's access to land and rights to plant trees on the site were rigorously controlled. While gardeners were invited onto the site for irrigation purposes, Lang Swaray was determined that their use rights would only be temporary. The landholder accepted no *kumakaa* payments for fence expenses, nor did he accept any assistance in fence maintenance or plot clearing; the irrigation of trees was the sole condition for garden access. The only trees that gardeners themselves were allowed to grow in the garden were papayas, and these were removed once the gardeners left the site. All other trees were owned by the landholder and were managed to produce a cash-crop income.[6] This strict adherence to self-financing and management was meant to ensure that no counter claims could be raised against the landholder when his trees matured and the women were asked to leave the garden.

By 1995, all of the gardeners in Lang Swaray's garden had been required to vacate the site. Several hundred improved mango and banana trees had begun bearing fruit, and the landholder had just used his own funds to install a new mechanical hand pump to assist with irrigation. In essence, Swaray had fulfilled all the major expectations of the Forestry Department's agroforestry extension agents. From the donor perspective, his was a site that profitably and sustainably integrated environmental and development goals. While ideal from the standpoint of the coalition of interests supporting agroforestry, the conversion of Swaray garden into an orchard had significant negative implications for the ninety gardeners who once held plots there. Despite the fact that the labor of these gardeners was instrumental in the establishment and eventual survival of the landholder's orchard, they derived none of the long-term benefits from the trees. All of the prime garden land along this stretch of Jowara Creek was either lost to the orchard or had already been taken by other gardeners. Moreover, landholders uphill from the gardens were unwilling to give the gardeners access to land primarily used for groundnut and millet production. Thus, the women from Swaray garden were left with no choice but to relocate their plots in swampland immediately adjacent to their old site. The move to the swamp had two principal effects. Since the new site had no fence, the gardeners were forced to patch together a "traditional" fence with bush posts, thorn bushes, and woven mats. This partial barrier did little to deter livestock from grazing on crops, however, and losses were substantial. More significantly, the gardens were located directly on top of rice plots that were flooded by seasonal rains. This meant that all vegetables had to be uprooted in favor of the rice crop each rainy season, thereby restricting the garden's productivity. While no data are available for Swaray garden itself, a survey conducted in 1989 in a community not far from Kerewan illustrates the potential cost of such a relocation. Garden plots in this second site, which had been relocated in swamps due to the closure of shade canopy over an upland site, yielded a weekly average of D136 ($18) per grower as late as the last two weeks of May in 1989 (Schroeder and Watts 1991).[7] The importance of this level of cash income—the rough equivalent of a bag of imported rice each week—during the pre-harvest period known as the "hungry season" is obvious. Yet the vegetable crop was summarily uprooted in order to resow the beds with rice. In effect, spatial limitations imposed by orchard development in the immediate uplands forced gardeners to choose between income generation and food production at a time of year when both objectives were indispensable.

While the closure of Lang Swaray's garden to vegetable production was a significant setback to the gardeners working on the north side of Kerewan, women working at two other sites had at least temporarily regained the upper hand in conflicts with their respective landholders. The first of these was the 3 ha garden / orchard project established by the Norwegian NGO described above, where project managers had vowed that gardeners working the site would be subject to rigid managerial control. When the Norwegian site was first established, it was laid out as a perfect rectangle in order to simplify the areal survey. The site's fence was constructed with expensive imported materials, including two courses of concrete blocks, heavy gauge chicken wire, and metal fence posts topped with barbed wire. Instead of skirting the adjacent older garden site, the new fence line ran directly through existing garden beds, whether the previous occupant had standing crops or not (fig. 13). As a result, the project erased women's property rights in an area favored by more elderly gardeners in particular because of its exceedingly shallow groundwater table.

By 1995, the sustainable income-generating project originally envisioned by project managers was in shambles. A symbol of the rigidity of that vision, the high-tech fence line, had all but collapsed. Of more than 400 mango trees planted by the agency, only 81 trees had survived, and less than half of these were in good health, this despite the project manager's rhetoric stipulating that trees on this site were to be protected at all cost ("the survival of trees is their first goal"). Some of these trees were no doubt lost to "natural" causes, such as disease or transplant "shock," but others were clearly the victims of concerted (in)action by the gardeners themselves. Physical inspection of the 81 remaining trees revealed at least 1 tree that had been deliberately girdled—its bark had been stripped by a gardener in order to hasten its demise. Other trees showed signs of acute drought stress from not being watered, or they had been damaged by livestock allowed to roam freely throughout the site after most of the vegetable harvest had been completed. Thus, the gardeners themselves contributed to the poor survival rate of the trees through willful neglect and outright sabotage.

It was telling that the manager who had been so adamantly opposed to gardeners in 1991 denied upon questioning in 1995 that the site was ever intended to be anything but a garden project for women. It seemed, in fact, that the agency had abandoned the confrontational approach it had originally espoused. Indeed, in the interim, the agency had hired as its new agroforestry manager the man who served as my principal research assistant during the early phases of this project (see Preface). And its new approach explicitly preserved gardeners' use rights by zoning garden / orchard spaces for multiple uses.[8] In the words of the new agroforestry coordinator, "Women will have their share there."

A third site that had undergone rapid changes in the early 1990s was Sanyang garden. There is a large mango tree at the center of Al-Haji Sanyang's site that graphically illustrates the struggles that mark the garden's history. In 1984, the big tree was one of the few mango trees spared when the landholder moved to purge the garden of women's trees. Between 1984 and 1991, it grew to be twenty or thirty feet tall and developed a large shade canopy that effectively precluded any cultivation under it. With this canopy in place the big mango reminded some gardeners of what the whole garden might have experienced had the landholder not stepped in to curtail tree planting when he did. Other growers with plots in the immediate vicinity of the tree put its shade to good use as a cool and safe haven where they could rest and socialize away from the beck and call of their husbands. In addition, the garden's radical leadership had claimed the area under the tree as an informal central meeting place where the day-to-day plans affecting the garden as well as more deliberate political strategies involving the group's membership were debated and developed. In other words, the tree stood for some as a symbol of landholder authority and paternalistic wisdom. For others, it was used as a day-to-day refuge from petty household-level tyrannies (and the midday heat), and for key members of the garden's leadership, it was a rallying point for more far-reaching collective action and decision making.

When I visited Sanyang garden in 1995, the big mango tree that towered over the center of the garden in 1991 looked as if someone had halved it with a meat cleaver. Several large sections of the tree had been removed entirely, and its once imposing profile had been dismantled, all of which indicated that some sort of fundamental shift in tenure regime had taken place. Gardeners in Sanyang garden had long been frustrated by their inability to solve their fencing problems. In the past, they had repeatedly complained that the landholder had taken money collected for fence maintenance (or well construction) on more than one occasion but had failed to make all the necessary repairs. Thus, in 1992, several leaders of the garden group approached the Save the Children Federation, which had since become active in the community, with a request that the agency provide them with fencing materials. Like other developers before them, SCF's field personnel were reluctant to step into an arena that had been the source of so much conflict through the years. But the gardeners were insistent. Their situation had grown increasingly desperate as their aging fence deteriorated. They could no longer keep goats, sheep, and cattle out of their fence perimeter, and this meant that their contributions to family livelihood were jeopardized.

In an effort to remedy the gardeners' dilemma and finally resolve the tensions between the garden group and the landholder, SCF brokered an agreement between all of the key parties to the dispute. The NGO committed itself to loaning the gardeners money to completely replace the old fence, but only if the landholder agreed to certain stipulations. The other landholders controlling parts of the site (see chap. 5) increased the pressure on Sanyang by threatening to subdivide the site if Sanyang continued his obstinate refusal to grant the women's wishes. In the end, with the full weight of the community's moral economy on his shoulders, Sanyang capitulated. The agreement he signed along with a representative of the garden group leadership was witnessed by the head of the Village Development Committee (VDC) and the district commissioner. It read in part as follows:

> This agreement is between [the landholder] and [leaders of the women's garden group] of Kerewan village, NBD and brokered by Save the Children. The landowner and the garden group agree to the following:
>
> 1. Al-Haji Sanyang will allow the group to continue gardening activities on the current site as long as the group wishes.
> 2. Sanyang will refrain from exercising ownership rights that may result [in] the hinderance [sic] or interference o[f] gardening activities on the said site.
> 3. The group has the right to plant any vegatables [sic] and fruits of their choice.
> 4. The group will not plant any trees that may interfere with gardening due to shading.
> 5. The group will receive a loan . . . to fence the entire garden with barbed wire. The group will pay back the loan in two years. . . .
> 6. The group will plant a live fence to protect the garden from animals.
> 7. Once the live fence is established and the garden no longer threatened by animals, the group has the right to decide whether to remove the barbed wire fence for other uses or not.
> 8. The group will continue to save a portion of their yearly earnings in order to repay the loan and maintain the garden with minimum outside assistance.
> 9. Should the landowner or the group wish to amend, in any way, or revoke, this agreement during the duration of the loan, they shall do so in consultation with Save the Children. Thereafter they shall do so in consultation with the undersigned witnesses.

What this agreement meant in practical terms was that gardeners in Sanyang garden were free not only to remove the big mango but to reconfigure the vegetable / tree crop mix at will. They had, in effect, finally acceded to a tenure status that accorded them untrammeled rights to control crop selection, site maintenance activities, and the direction of future development interventions. In exchange, the agency won an agreement on the part of the gardeners that they would work to replace the old wire fence with live fencing, thereby achieving a limited environmental objective and, theoretically, removing one of the key sources of conflict on the site.

Thus, after nearly two decades of struggle, Sanyang garden became for all intents and purposes a "women's garden." Unfortunately, this victory was muted somewhat by events at the national and international political level. In 1996, Save the Children, the agency that brokered the new agreement, decided to leave the country, in part because some of the agency's key funding sources, donors such as USAID, were withholding aid from The Gambia following the 1994 military coup.[9] Whether this turn of events left gardeners in Sanyang garden vulnerable to some new challenge by the landholder remains to be seen. As for the landholder, having finally relented in the battle over rights to his former site, he simply relocated to one of the few remaining stretches of undeveloped low-lying land south of town, where he established a new fence perimeter under his name. Following along the fence at the new site, planted squarely within garden plots allocated to women, was a line of mango seedlings, each belonging to Al-Haji Sanyang.

AGROFORESTRY INTERESTS

What the various illustrations cited in this chapter demonstrate is that the politics of agroforestry in Kerewan's garden districts were far from overdetermined by any single set of causative factors. Each of the groups of actors involved in the development of low-lying land along the North Bank brought a critical component to the task of reclaiming these lands for productive purposes: landholders controlled the land itself, women gardeners their own skills and labor, and development agencies capital and material inputs. In retrospect, it appears that neither garden projects nor orchard initiatives were very successful without tapping into all three resource pools simultaneously. The negotiations between the three groups were accordingly protracted, and in the end they became quite unpredictable, an outcome that was due as much as anything to the somewhat fickle approaches adopted by the developers themselves. A systematic review of the competing sets of interests held by gardeners, landholders, and developers vis-à-vis agroforestry is thus necessary in order to fully assess the political ecological fallout of the agroforestry promotional effort.

In the initial period of the garden boom, women were granted usufruct rights to low-lying land and, with them, a rare, if still somewhat limited, opportunity to attempt long-range land developments such as tree planting. They were also presented with favorable material circumstances: outside investment in irrigation infrastructure and the creation of secure fence perimeters made it possible to grow trees more intensively than ever before. Thus, hundreds of Kerewan women planted trees in their garden plots, both as a source of additional income and as a way to produce micro-environmental conditions more favorable to vegetable production (Barrett and Browne 1991; Mann 1990). In the mid-1980s, as new market outlets opened and new seed varieties became available, the economic calculus changed in favor of vegetable production. Moreover, income from vegetable sales gave many women the necessary leverage to gain for themselves considerable new social freedoms. Gardeners responded to these changes by cutting back trees or chopping them down in order to open up the shade canopy and expose their vegetables to sunlight.

In this context, trees grown by landholders represented a special threat. For with landholders becoming more deeply involved in tree crop production, the nature of the agroforestry system itself changed significantly. It was no longer a system that allowed the women to unilaterally diversify their crop selection and spread financial risk but a *successional* system deliberately designed to bring about a transfer of use rights and effective control from women gardeners to male orchard owners. The agroforestry system promoted by developers and adopted by landholders threatened to undermine the garden-based livelihood system, and gardeners consequently resisted it through aggressive tree trimming, "accidental" damage to trees by fire, and malign neglect of watering responsibilities imposed upon them by landholders.

The men who intervened to plant trees in gardens were motivated by many of the same factors that led to the garden boom in the first place. Just as was the case with WID projects and women farmers before, would-be orchard owners sought to take advantage of a favorable shift in development ideology and practice, albeit from the powerful position vested in landholding status. Landholders were particularly interested in alienating garden subsidies. Garden projects increased the value of low-lying land immensely through the addition of concrete wells and wire fences, and the extensive soil improvements and "local" wells provided by individual gardeners. As Thoma notes: "the attraction of free wells, subsidized fencing, subsidized seedlings and the like [was] virtually too irresistible . . . not to take advantage of, even if [male] farmers already [knew] how to plant and grow trees . . . on their own" (Thoma 1989, 41). In addition, the labor source available to landholders in the gardens had become more secure. Having invested hundreds of hours of personal labor in extensive site improvements, gardeners had become more "rooted" in place in their individual plots. This allowed landholders to "capture" the women's labor for their own purposes.

The agroforestry initiatives embraced two different forms of tenure claim advanced by landholders, one focused on existing sites, and the other on new sites that they themselves established. In existing sites, the location of trees directly on top of garden plots threatened to force the women off the land and alienate the numerous improvements they had made to garden sites. On new sites, women were not directly dispossessed of previously held rights, but they were made to forfeit the privileged position they had at least temporarily enjoyed in determining the disposition of development aid. Thus, in one case landholders reclaimed land directly from gardeners, and in the other, they reclaimed practical control over land vested in the ability to leverage outside resources toward its development (cf. Ribot 1995).

In addition to income generation, men used tree planting in gardens as a way to regain lost social prestige and overcome perceived threats to the lineage landholding system. As noted above, the generally poor groundnut market left male farmers searching for ways to diversify their economic activities. They felt the surge in female incomes created an imbalance in household social relations that needed to be redressed. According to this perspective, women gardeners were said to lack respect for their husbands, and their newfound wealth had placed them in the position of being able to buy themselves out of bad marriages altogether. Tree planting accordingly had a double significance for landholders: it was a means of taking advantage of shifting development priorities; and it constituted a reassertion of male landholding privilege. In this context, all investments, whether from WID programs or forestry projects, were welcome, since landholders stood to benefit from them in the long run anyway. The key was the enclosure and bounding of female labor resources that made private orchard development possible.

Perhaps the most difficult set of interests to explain were those that motivated the various development agencies involved in market gardens and agroforestry projects. In 1987, a team of environmental experts toured West Africa on a mission set by the U.S. Congress. Their mandate was to find and retrieve "success stories" in natural resource management on the Sahel (Shaikh et al. 1988). Congress had grown weary of African disaster narratives and wanted to determine what successes, if any, had been achieved before allocating additional funds toward improving natural resource management in the region. By the time the team reached The Gambia's North Bank, they had been on the road for weeks and were nearing the end of the Gambian leg of their research trip. It was my task as coordinator of Save the Children's agricultural and natural resource management efforts to provide the group with a tour of Save the Children's projects. As I sketched out the day's proposed itinerary, one of the leaders of the group commented, "I don't care what we see, just don't show us another garden project." He elaborated that at virtually every stop over the course of the team's two- or three-day Gambian investigation, they had been shown trees planted within communal gardens as evidence of the host agencies' commitment to environmental stabilization goals.

That this group of specialists should come away from their trip through The Gambia with such an impression of the nature and scope of development efforts related to natural resource management in the country is really quite remarkable. This is especially so given the limited impact garden / orchards have had on the problems of deforestation in the country. One survey of land use patterns in The Gambia, for example, grouped gardens with housing, roads, and other "communal" areas in a single land use category, yet these "communal" uses still only accounted for 2 percent or less of all land area in the villages surveyed (Thoma and Jaiteh 1991). The fact that so many of the development agencies hosting the traveling team of experts chose to base their environmental credentials, as it were, on such tiny garden / orchard projects begs the question of what really lay behind the move to create the projects in the first place.

Developers working in Africa in the 1980s and early 1990s came under heavy pressure from their various constituencies to demonstrate that they were doing something to quell the "eco-disaster" looming on the continent (Barker 1989; cf. Watts 1989). The prominent display of images of drought and famine in American and European media made environmental initiatives the sine qua non of development in the African region. It was for this reason that the relatively tiny, oasis-like women's gardens assumed larger-than-life proportions on the development landscape. Reconstructed by developers as sustainable agroforestry projects, they offered up something of a mirage of ideological commitment, action, and success on an otherwise fairly bleak horizon and came to form the basis of the developers' institutional legitimacy vis-à-vis their donors.

These developments constituted no mirage for vegetable growers, however. Instead, the coherence of policy statements and practice marked the formation of a new political alliance in The Gambia's garden districts. Whereas the coalition confronting landholders in the 1970s and early 1980s sought to shore up the property claims of fledgling horticultural groups vis-à-vis the entrenched interests of the lineage-based landholding system, the late 1980s and 1990s saw many of the same agencies effectively switch sides and endorse landholders in their efforts to reestablish prerogatives over garden plots and implement plans for income generation in the service of environmental objectives. To be sure, gardeners resisted and even overcame some of these constraints, but the intentions of the new pro-orchard alliance were clear. It may be worth digressing a bit to consider how and why this might be the case.

AGROFORESTRY THEORY

Agroforestry systems have been widely touted for their prodigious capacities.[10] In the context of the global ecological politics of the 1980s and 1990s, where commercial interests ranging from green marketeers selling rainforest crunch to biotechnology prospectors mining tribal gene pools have exerted enormous influence (Smith 1996), agroforestry approaches have accomplished dual purposes. They have simultaneously boosted commodity production and contributed toward efforts to stabilize the underlying resource base. On both of these grounds, agroforestry approaches have been constructed as an unambiguous and unalloyed "good" (Rocheleau and Ross 1995; cf. Schroeder 1995). Institutional actors in forestry and environmental agencies and the major multilateral donor agencies such as the World Bank have accordingly joined forces to promote and preserve agroforestry in many parts of the world.

Several rationales have been developed in support of this effort. From a production standpoint, intercropping trees with other crops can fix nitrogen and improve nutrient cycling, enhance chemical and physical soil properties, conserve moisture, and make generally efficient use of a range of limited yield factors (Altieri 1987; Anderson 1990; Nair 1989). Similarly, from an environmental standpoint, agroforestry systems may reduce erosion, provide alternate habitat for wildlife, and shelter a diverse range of plants and other species (Redford and Padoch 1992). These systems have thus been of equal concern to agricultural economists interested in maximizing returns to land and labor inputs and biologists and ecologists interested in environmental preservation and protection. Both of these aspects appealed to development agents promoting such systems in The Gambia.

While the attributes of agroforestry systems that contribute toward production and conservation goals have been paramount in the eyes of many development donors, the virtues of agroforestry systems have also been praised by scholars and activists interested in restoring or protecting the rights of otherwise disenfranchised peoples. Rocheleau (1987) demonstrates quite clearly how women have mobilized agroforestry strategies to make the best use of the minimal landholdings typically allotted to them in most land tenure systems. Fence lines, backyards, roadsides, hedgerows, ravines, the understory of commercial tree crops—the marginal "in-between spaces" located in the interstices of tenure systems otherwise oriented almost exclusively to the needs of men—these are the sites that women have colonized and rendered productive through agroforestry techniques (Fortmann and Rocheleau 1985; Rocheleau 1987, 1995; Rocheleau et al. 1995; and see also Leach 1992; Madge 1995b). The opportunities to mix crops inherent in many agroforestry systems—food crops with cash crops, men's crops with women's crops, fuel and fiber crops with fodder and medicinal crops—have been especially important when other opportunities have been foreclosed, or when, as has often been the case with women farmers, time and energy constraints have made efficient collection and gathering of resources a necessity.

Other authors have noted that agroforestry techniques have been widely used among indigenous peoples in many parts of the world (Brush and Stabinsky 1996; Clay 1988; Redford and Padoch 1992; Zerner forthcoming). In this context, they have formed integral components of complicated, multiple-use resource management strategies that integrate crop production with hunting, gathering, and specific ritual activities. They have also served in many ways as repositories of the environmental knowledge indigenous groups hold. As such, they may yet provide keys to understanding how particular ecosystems were originally produced and, by extension, how they must be maintained if the biological and genetic resources they contain are to be conserved. From the standpoint shared by many indigenous peoples, there is much more at stake than just conservation issues, however. Agroforestry systems may reflect knowledge that is vital to outsiders, but they have also provided the means for generating local livelihoods and have been essential in many ways to cultural identity. Thus, while conservationists have sought to learn more about agroforestry and other resource management systems deployed by indigenous groups, the issues of intellectual property rights and just compensation for the sharing of local knowledge, the lack of viable economic alternatives, and, ultimately, cultural survival have remained critical for many indigenous groups themselves.

Dove's (1990) analysis of the diversity and complexity of "home garden" agroforestry systems in Indonesia invokes a third set of social relations involving the peasantry and the state. At stake is the peasantry's ability to resist the state's efforts to extract surplus through taxation and various forms of unequal exchange. State agents have monitored closely the production of rice and other commodities bound for external markets such as teak wood (Peluso 1992), but they have had relatively little interest in, or knowledge of, many of the other resources peasants produce through home garden-based agroforestry practices (see also Fried forthcoming; Michon et al. forthcoming). In this context, the fact that home gardens have incorporated a wide range of plants with high use value but low exchange value is key. Agroforestry techniques have thus been instrumental in helping rural peasant groups resist petty state tyranny.

Each of these perspectives, Rocheleau's outlook on gendered tenure, the concern of cultural survival advocates for indigenous peoples' rights, and Dove's emphasis on peasant resistance strategies, makes the same basic point regarding the social, economic, and cultural significance of agroforestry systems for specific groups. They draw attention to the fact that agroforestry systems have frequently been sites of contentious political struggle. Conflicts over agroforestry practices on The Gambia's North Bank suggest that this point can be more finely drawn, that two dynamics in particular contribute toward heightening the political sensitivity of agroforestry systems. Property and labor arrangements seem especially fraught with political tension in "successional" agroforestry systems and in systems that are explicitly geared toward commodity production.

Successional agroforestry is premised on the underlying agricultural crop being gradually overtaken by trees as the system matures. What is often overlooked in its promotion is the fact that the planned succession of species may also imply a succession of property claims, with one set of interests eventually superseding all other claims. An oft-cited model for this approach is the *taungya* forestry system invented by British foresters in teak-growing districts in colonial Burma (Bryant 1994).[11] This agroforestry arrangement granted peasant farmers temporary land use rights in large-scale teak plantations owned and managed by the state. In theory, peasants were allowed to plant food crops within the plantation for a limited period of time in exchange for contributing their labor toward planting and maintaining tree seedlings. In practice, peasants often sabotaged the seedlings through arson and neglect in order to maintain access to land for agricultural purposes. Thus, while this system has been used to argue that highly rationalized planning methods can ensure sustainable use, it has rarely been a complete success on its own terms, given the resistance it has engendered. In fact, as Bryant (1994) argues, the *taungya* system developed in Burma was not the pure product of "scientific" planning at all but was created by default when colonial foresters were unable to establish or maintain teak plantations by any other means due to peasant opposition. The *taungya* system was thus first and foremost "the outcome of an antagonistic relationship between an acquisitive colonial power and a threatened indigenous people whose reactions varied from covert resistance to defensive compliance" (Bryant 1994, 21). Any attempt to reproduce successional agroforestry systems along these same lines "remains an invitation to popular sabotage" (Bryant 1994, 25). The evidence from The Gambia suggests that women gardeners acted on this invitation repeatedly.

If the political nature of many agroforestry systems is evident in successional systems, it is also transparent where agroforestry approaches are heavily commercialized. Such systems tend to extend and rigidify (Raintree 1987) the tenurial rights of tree growers vis-à-vis competing resource users, such as cultivators of underlying crops, forest product collectors, and pastoralists. As Bryant's analysis of *taungya* suggests, the commodification of tree cropping can drive a wedge between holders of tree and land or crop rights, and this polarization may in turn produce a range of agro-ecological and social contradictions. Moreover, the commercialization of agroforestry regimes often erodes moral economies and replaces them with a morally indifferent (not to say bankrupt) stance that elevates profit taking above all other objectives, including ecological stability (Schroeder 1995). With such social dynamics embedded in combinations of tree and understory crops, the design and implementation of agroforestry systems and the responses of different social actors operating within those systems must be carefully analyzed.

At a minimum, there is a need to move beyond technocratic and managerial classification systems (Farrell 1987; Nair 1989, 1990) to distinguish between agroforestry systems on political-ecological grounds. Systems such as those described by Dove, the cultural survival advocates, Rocheleau, and others often embody culturally diverse knowledge systems and essential livelihood practices. They are thus fundamentally different in scope and purpose than contemporary strategies pressed into being by forest managers, environmental NGOs, and economic developers bent on merging environmental and commodity production objectives. There is a striking contrast between systems that actually accentuate and preserve a diversity of species, uses, and claims and those that practically narrow the range of options within each of these parameters.

In sum, my extended analysis of the Kerewan case suggests the usefulness of monitoring the specific forms of environmental intervention that have taken shape over the past two decades. For each form carries with it its own pattern of opportunities and constraints, its own signature set of property dynamics and characteristic mechanisms for labor control. I have acknowledged above that agroforestry systems can constitute a key site of refuge for women working to overcome the constraints of male-dominated land tenure systems, indigenous peoples struggling to preserve cultural identity and sovereignty over ancestral lands, and peasantries resisting the extractive propensities of the state. These are all opportunities for otherwise disenfranchised groups that are worthy of defense. By contrast, the successional agroforestry system introduced in The Gambia was a means for male landholders to reclaim their control over low-lying land and, by extension, undermine many of the social gains their wives and female relatives had made at the landholders' expense. While these controls were successfully contested by gardeners in many cases, they have added to the difficulty of securing livelihoods through the market garden system. These practices clearly verify Watts' argument that: "[r]ights over resources such as land or crops are inseparable from, indeed are isomorphic with, rights over people" (Watts 1993, 161). Thus, it is a particular *form* of agroforestry, the planned transition from multi-cropping and shared benefits to mono-cropping and privatization, that remains so problematic and bears such careful scrutiny (cf. Bryant 1994; Peluso 1992; Suryanata 1994).

CHAPTER 7

Shady Practices

DEBATING DEVELOPMENT REFORM

In this book I have explored successive attempts to reclaim low-lying soil and water resources in a small town on the North Bank of the River Gambia. In the first instance, rural women's groups, responding to drought-related changes in key agro-ecologies and a squeeze on household finances engineered by the promoters of structural adjustment embarked on a venture to intensify production of fresh vegetables and fruit for commercial sale. As these women gradually improved their horticultural knowledge and practices and forged crucial market connections, women and sympathetic male colleagues in the United States and Europe, the centers of development capital, began working their way into international donor agencies, where they fought to influence hiring decisions and redirect development program objectives toward women's needs. In the 1980s, several NGOs, voluntary agencies, and mission groups established themselves in The Gambia. Armed with capital and a commitment to "bring women into development," these groups were immediately attracted to the increasingly viable horticultural enterprises. Hundreds of small grants for garden wells and fences were quickly negotiated, and a veritable boom in market gardening ensued.

As these developments took place, a second set of initiatives to reclaim lowlands through tree planting slowly emerged. Male landholders, who had recently seen their groundnut-growing enterprises all but collapse under the weight of poor producer prices, high costs for inputs, and drought-induced yield reductions, began experimenting with new mango varieties in hopes of establishing fruit orchards. Like the market gardeners before them, the orchard owners benefited from a favorable shift in development policies and practice, as several agencies that had once supported the garden boom adopted new objectives centered on the tasks of producing a biologically diverse landscape and recreating the conditions necessary for sustained food security. Again, hundreds of small projects were initiated and the orchards became a significant addition to the rural landscape.

In the eyes of their promoters, these developments could not have been more timely or appropriate. The gardens represented a substantial diversification of the agrarian economy and gave rural families a much needed financial boost at a time of serious economic difficulties. Tree planting, in turn, was seen as an essential strategy to reverse degradation of the country's natural resource base, an approach that also generated a recurrent source of income in the form of fruit commodities. Thus, both gardeners and orchard owners were at least partially successful in making environmental management profitable. In this regard, they achieved the ultimate goal of donors who sought to "integrate" environment and development through productive endeavors (Biodiversity Support Program 1993; Brown and Wyckoff-Baird 1992; USAID 1993; World Bank 1996).

While these representations of Gambian gardens and orchard projects contain several essential truths, they have nonetheless grossly simplified and distorted the lived realities of North Bank residents and the agencies that serve them. Such promotional tendencies are widespread in the development literature, and for that fact alone, it is important to try to understand how and why they occur. The notion originating in feminist and environmental critiques that wealthy development donors and project implementing agencies needed to address questions of gendered power relations and confront the environmental consequences of prevailing development practices gave rise to a complex political process. Initially, bureaucrats in many agencies sought to blunt criticism of their efforts by simply denying that the alleged problems existed. When their detractors persisted, however, donors were forced to acknowledge responsibility and accept blame for their failures and ultimately began to develop programs that at least nominally addressed gender and environmental concerns.

The question of whether these reforms in fact represented genuine steps forward has provoked sharp differences of opinion and occasionally divisive debates within both the environmental and feminist communities (*The Ecologist* 1993; Jackson 1993; Merchant 1992; Sachs 1993). Some scholars and advocates have seen the donors' involvement in sustainable development and gender programming as a major breakthrough, and they have dedicated their own advocacy, consulting, and academic practices to furthering these gains. Others have remained more skeptical, interpreting the gender and sustainable development initiatives as forms of co-optation that only justify further condemnation.

My own conclusion based on evidence drawn from The Gambia and elsewhere is that the political character of much of what passes for gender and environmental programming is highly ambiguous. At a theoretical level, these, the eponymous shady practices, entail a great deal of conceptual slippage. There is a vast difference, I would argue, between the precepts offered by the feminist environmentalists and political ecologists cited in chapter 1, concerning women's work and women's rights, and those underlying the NGO declaration that Gambian women should be cherished as the country's "most precious and vital local resource." Indeed, the summary impression from the Gambian case is that rationales ostensibly intended to bolster programming more sensitive to issues of gender and the environment have in fact been turned against women gardeners (and their supporters) in practice. Simple rights once held by gardeners such as the selection of crops to be grown on a given location or the decision to trim a tree branch shading out a garden plot, as well as more fundamental privileges such as access to land, water, and labor have been challenged by environmentally conscious developers who have asserted the value of tree crops over all other objectives and worked to capture women's labor to help meet environmental goals. To be sure, the gardeners' implicit response to these ideas has been to thwart their application at nearly every turn. They have used arson, sabotage, and malign neglect to combat tree planting whenever the intentions of landholders and extension agents ran contrary to their own. But this realization does little to alter the fact that much of the original political intent of gender and environmental critiques has been lost as the major development agencies have adopted gender and environmental principles as their own.

These reflections suggest two directions for further inquiry. There is a need to amplify existing analyses of development in order to more systematically explore the ways in which development planners co-opt critical ideas. James Ferguson (1990) argues that the development apparatus works to systematically depoliticize economic, social, and cultural changes that accompany development interventions. His devastating critique of World Bank and other donor policies in Lesotho shows convincingly how problem definitions contained in development plans presuppose solutions that only development technocrats can offer. This leads to the expansion of bureaucratic power at the expense of local sovereignty, regardless of whether a given program's explicit objectives are met. While these dynamics have certainly played a role in The Gambia, the evidence drawn from the Gambian case suggests that they are only part of the story. The different ways of rationalizing the use of Gambian women's labor in agroforestry projects show that the donor and implementing agencies have been less interested in neutralizing their critics through bureaucratic inertia than in absorbing criticisms and reformulating them. Rather than simply downplaying politics, they have reinscribed critical discourses with new political meanings. Thus it was, for example, that trenchant feminist critiques in the late 1970s were translated into reformist WID objectives and radical environmental demands in the 1980s became the bland and often contradictory policy prescriptions proffered by many proponents of sustainable development. Both of these moves left the basic premises of the development enterprise intact. In this regard, while the means have differed, the effects of donor efforts have been similar to those described by Ferguson. By stripping the critiques of much of their original political content, the major donors have felt free to embrace gender and environmental objectives as part of a new political platform and have accordingly claimed a commitment to progressive ideals without significantly altering many of their basic practices.

A second direction for further inquiry is more reflexive in nature, for it must be acknowledged that at least some of the problematic assumptions adopted by developers in The Gambia originated with the critics themselves. It is striking to note how quickly the debate over gender and the environment detailed in chapter 1 manifested itself in the material practices of developers working in The Gambia River Basin. The evidence drawn from planning documents and reports shows quite clearly how ecofeminist and feminist environmentalist views were reflected in policies designed to incorporate women into natural resource management. When the WID focus of the 1970s and 1980s gave way to policies reflecting the influence of ecofeminist and feminist environmentalist critiques in the 1990s, many of the gains made by market gardeners during the boom years were threatened. Instead of supporting women in their efforts to expand land use rights, some developers used the idea of a naturalized connection between women and nature and the double-edged notion of women as environmental victims / managers to justify basing rural development policies on the premise of using unpaid female labor. These and related efforts by the major development agencies to "mainstream" (World Bank 1995) gender and environmental concerns challenge advocates to come to grips with the prospect that the uncritical application of their ideas may have serious, if unintended, negative consequences. The potential for co-optation of critical ideas stands as a reminder of the need to continually hone theoretical insights, a caution that holds equally, and perhaps especially, true in "applied" contexts.

QUESTIONS OF POWER AND JUSTICE

The interpretive challenges presented by the Gambian case study do not end with the problems of assessing developer motives and practices and the difficulties of sorting out a coherent position in the theoretical and political debates they have engendered. The question remains how to assess the outcome of the gender dynamics that took shape over the past two decades along The Gambia River Basin. It should be quite clear from the discussion in this book that women in the North Bank garden districts won significant gains for themselves over this time period. In personal terms, the gardens provided incomes that allowed gardeners considerable discretion with regard to their personal needs. They placed women in the position to buy their own clothes, pay for their own travel, and send their own children to school, all without suffering the indignity of asking their husbands for money. Thus, garden proceeds became a tool for women to buy freedom of movement and broker a new conjugal contract, to win for themselves "second husbands." The gains on domestic fronts were mirrored at the point of production where women used tree planting, surreptitious land transfers, and leverage provided by their allies in the development agencies to extend their usufruct rights unilaterally. They then managed to ward off some of the most egregious attempts by landholders to reclaim garden lands, again with assistance from NGO personnel who intervened to help resolve what they perceived to be natural resource management disputes.

In broader social terms, garden projects also had profound effects on local livelihoods. During the early stages of the garden boom, rural Gambia was staggered by the double onslaught of drought and structural adjustment. After the poor rainfall years of 1982–1984, rural fam-

ilies in The Gambia had difficulty coming up with sufficient *seed* to plant
a new crop, much less meet their families' needs for grain staples the
year round. At the same time, thanks to structural adjustment, subsidies
on fertilizer prices were eliminated and the costs of producing other cash
crops soared, even as food prices doubled overnight. The gardens were
thus as key to household reproduction as they were to individual strategies of accumulation, and women worked hard first to calibrate and ra-
tionalize their productive and reproductive labor obligations and then
to resist alternate plans for development of the garden sites with whatever means they had at their disposal. The net effect, in the words of one
gardener's husband quoted above, was that by the mid-1980s, thousands of rural households were "using gardens to survive."

In sum, it is clear that international gender critiques helped produce
real gains in rural Gambian women's incomes and enhanced their collective power and prestige. Moreover, the intervention by outside developers into the land politics of the garden boom created opportunities
that women gardeners shrewdly exploited. Gardeners assumed numerous rights and privileges over *boraa banko* lineage land, often without
the knowledge of, and at times in the face of direct opposition by, male
landholders. The fickle nature of developers' interventions sometimes
aided them in this process, as they struggled on a site-to-site basis to
maintain their gardens in the face of intense pressure.

To acknowledge these gains is not the same as denying the negative
impact of the agroforestry projects, however, for critical questions of
power and justice remain unresolved. It is one thing to concede that
women as well as men can and do "play" the game of development, that
they, too, act on the basis of motives that are narrow and mean at least
as often as they aspire to enlightenment, and that they sometimes win
the struggles over land, labor, and livelihoods initiated by development
interventions. It is quite another to argue that the structural determi-
nants operating within Gambian social systems no longer have any
force, or that the development interventions designed to incorporate
women into environmental management have not produced deleterious
effects in many areas.

Some analysts confronted with the data from The Gambia's garden
boom have argued that the male landholders in the garden districts
should not be faulted for planting trees in gardens because they "owned"
the land in question and therefore had the right to develop it as they saw
fit. However, this contention is based on two erroneous and damaging
assumptions: first, that the rights accruing to landholders in the lineage

system are exclusive rights akin to private property holdings in a capitalist system; and, second, that claims to land are inherently stronger than, and thus supersede, claims to one's own labor. These arguments are both steeped in neoclassical economic theory, and as such, they almost inevitably overlook the most salient social facts of the case. The latter contention is particularly insidious when viewed from the perspective of feminist political ecology, since it is precisely property in land that women in many social geographical settings lack (cf. Rocheleau et al. 1996). If the default assumption is that land claims are always prior to labor claims, women have few prospects indeed of ever aspiring to improve their individual and collective circumstances. In practice, the landholding rights a select group of men hold in Mandinka society are more properly understood as *stewardship* rights rather than private property rights, and the relative weights of claims over land and claims to labor are, if not commensurate, than certainly the product of contestation (cf. Carney 1988a, b). They must accordingly be seen as historically and geographically contingent—as opposed to natural—conditions on an ever-changing social landscape. The basic fact remains that most orchard projects in The Gambia have been based on the use of unpaid female labor. If the assumption that land claims have priority over labor claims is reversed, then the use of women's labor on orchards has no inherent moral basis whatsoever. There are clearly mechanisms for sharing the social product of orchards in Mandinka society, and where these have been carefully negotiated, amicable solutions have been worked out between landholders, women's groups, and donors and orchard projects have flourished. These cases notwithstanding, the attempt to usurp women's labor in the service of broad environmental objectives without their explicit consent simply cannot be justified.

The subjugations of the marriage system, the inheritance system, the landholding system, and so on, continue to present obstacles to the exercise of basic rights and privileges by Gambian women. Indeed, they remain the very seat of patriarchal authority and should not be neglected in the rush to claim victory for women market gardeners. The agroforestry interventions designed and implemented in the Kerewan area have only bolstered the powerful positions an elite group of senior men have enjoyed by re-essentializing market gardeners as workers naturally tied to the landscape and as mothers who forfeit everything in the service of their families. These practices are clearly inconsistent with even modest gender and environmental reforms and should be resisted.

Notes

PREFACE

1. Tensions between men and women in Alkalikunda clearly predated SCF's arrival. In the year prior to the garden dispute, the community's men had cleared a bush path to a swamp location several kilometers away under the assumption that the women would shift their rice production activities to that area. The women objected to the new site, however, arguing that it was too distant and in an area that would require constant surveillance to ward off damage from bush pigs. In the end, the women simply refused to cultivate the new area, saying, "You men have built that road for the baboons, because we will never use it!"

2. "The Gambian household [*kordaa*] is referred to as a 'compound.' In physical terms, the compound is the structure that serves as the arena for family interactions. In this sense, the compound constitutes a collection of dwellings, kitchens, and stores that is clustered around a central courtyard and fenced-off from the pathways outside. The number of people in one compound may exceed 100; the national average stands at around 16. These . . . people . . . share various rights, duties, and material possessions, . . . [and typically] recognize the overall authority of a single head (usually a male). . . . In large compounds a subdivisioning of distinct spheres of responsibility and activity becomes apparent. The decision-making units in both production and consumption are multifaceted" (von Braun and Webb 1987, 3). Production among the male members of the household is organized into units known as *dabadaalu*, whereas female production units *(sinkiroolu)* are smaller in scope and are often coterminus with cooking and consumption units (Dey n.d.; Gamble 1955). Garden production units are described in greater detail in chapter 4.

3. Forty of these women were selected from one of the town's largest gardens to the south of town, forty from an equally large site to the west of town, and

twenty from smaller, newer sites to the north of town. Each group was selected following a stratified random sampling method: every fifth woman encountered along a systematic walking path through the garden was incorporated into the sample until the desired number for each site was reached. In terms of the research design, this pattern was intended to capture some of the different production dynamics faced by gardeners in different locations surrounding the community, as well as the differences embodied in individual site histories. As the research progressed, it became evident that, in fact, most women had gardens in more than one site. Thus, while it was still possible to analyze particular dynamics on a site-by-site basis, conclusions were not forthcoming about the fortunes of individual women who sometimes worked simultaneously within multiple tenure domains.

4. The classification of domestic units into one of three economic categories was based on a set of crude economic indicators (e.g., possession of farm equipment and draught animals, construction materials used in residential buildings, off-farm employment, ownership of luxury items such as radio / tape players and gold jewelry). This information was subsequently supplemented with data on access to labor resources and other economic assets (e.g., economic trees owned by family members) for selected households.

5. One woman in the garden income sample left town before the production survey was implemented. Thus, the second survey's sample size was only ninety-nine.

6. Following the expansion of two sites in 1991, this figure grew to 22.5 ha. See table 14 below.

7. Thanks to Cindi Katz for suggestions along these lines.

8. The exact composition of my research team varied from task to task. I employed a team of six enumerators and two coordinators for the community census. Five members of this team were men and three, including the overall coordinator, were women. I also worked with a group of male enumerators on the physical survey of the gardens and on certain aspects of the income survey.

CHAPTER 1.   INTRODUCTION

1. President Jawara was replaced by a military council following a coup in 1994. The Gambia's second president, Yahya Jammeh, one of the members of that council, took office in 1996 following an election that was deemed unfree and unfair by international observers.

2. As of 1991, there was only a single male gardener in the community.

3. See Braidotti et al. (1994), Kabeer (1994), and Marchand and Parpart (1995) for more extended discussions of this history.

4. Among the agencies that adopted WID as a formal programmatic focus early were SIDA (the Swedish International Development Agency—early 1960s), and the Ford Foundation (early 1970s). USAID (1979) and CIDA (Canadian International Development Agency—1984) made WID a major programmatic emphasis shortly after the Mexico City meetings. By contrast, while some of the larger multilateral agencies had WID advisors in the 1970s, WID only became formalized as official agency policy at UNDP (1986) and the World Bank (1987) a decade later (Chowdhury 1995; Rathgeber 1995).

5. Much of this material is unpublished. See discussions in Bruce and Dwyer (1988), Kabeer (1994), and Whitehead (1981). The high percentage of female-headed households in many countries is also relevant in this context (Fortmann and Rocheleau 1985).

6. The origins of this argument lie in Ortner's essay, "Is Female to Male as Nature Is to Culture?" (Ortner 1974). Margarita Arias articulates the basic premise quite forcefully: "There is . . . no more appropriate place for women than as defenders of the planet. No one speaks out for the protection of the environment with greater moral authority than women. Only those who have fought for the right to protect their own bodies from abuse can truly understand the rape and plunder of our forests, rivers and soils" (Arias 1992, 75).

7. According to Rocheleau, Thomas-Slayter, and Wangari, "environmentalists" are "liberal feminists" who "have begun to . . . treat women as both participants and partners in environmental protection and conservation programs" (Rocheleau et al. 1996, 4). As a programmatic focus, this goal bears striking resemblance to the original WID objective of bringing women "into" development.

8. See Plumwood (1993) for a vigorous critique of the idea that such a set of "core ecofeminist principles" exists.

9. Rocheleau, Thomas-Slayter, and Wangari define "socialist feminism" and "feminist post-structuralism" as follows: "*Socialist feminists* have focused on the incorporation of gender into political economy, using concepts of production and reproduction to delineate men's and women's roles in economic systems. They identify both women and environment with reproductive roles in economies of uneven development . . . and take issue with ecofeminists over biologically based portrayals of women as nurturers. . . . *Feminist poststructuralists* explain gendered experience of environment as a manifestation of situated knowledges that are shaped by many dimensions of identity and difference, including gender, race, class, ethnicity, and age. . . . This perspective is informed by feminist critiques of science . . . as well as poststructural critiques of development . . . and embraces complexity to clarify the relation between gender, environment and development" (Rocheleau et al. 1996, 4).

10. As Watts (1983a, b) and others have repeatedly argued, drought does not automatically lead to famine. During the whole of the 1972–1974 famine, during which Ethiopia lost a quarter of a million people, it remained a net exporter of food; in 1984–1985, it exported vegetables to Europe in large quantities (Kebbede and Jacob 1988).

11. Rainfall totals rebounded across most of the region in 1969, but most communities were unable to regroup sufficiently in a single year for the brief reprieve to matter. The cumulative effect of drought over the next three years was devastating.

12. The cycle of droughts continued in the 1990s. A second major drought episode hit Southern Africa in 1992: "Southern Africa is experiencing its worst drought in living memory. Cereal production has been devastated in Zimbabwe, Lesotho, Swaziland, Malawi, Mozambique, Zambia and parts of South Africa and 18 million people are threatened with starvation. The region has lost a larger proportion of its crops than Ethiopia and Sudan in the devastating 1985

drought. The impact has been exacerbated by the fact that the two most abundant food producers, Zimbabwe and South Africa, are among the worst hit and are no longer able to export surpluses" (MacRae 1992; see also Meldrum 1992).

13. "Bulk and bagged cereals are critical commodities in most cases and their passage from overseas ports to main distribution centres in drought-affected countries may encounter several bottle-necks—from ship to pier; on quays and in port sheds; from port to main distribution centres; and when crossing national borders. Poor road networks, lack of repair and maintenance, inadequate spare parts, shortage of vehicles and storage facilities pose obstacles" (United Nations 1984b).

14. The arch-conservative economic ideologies of Ronald Reagan (1980–88) and Margaret Thatcher (1979–90) facilitated this process. The Reagan-Thatcher mandate consisted of a renewed emphasis on privatized solutions to development dilemmas. Self-help, often in the guise of entrepreneurism, was widely promoted throughout the region (Watts 1994). At the same time, state-level bureaucracies were decimated through "structural adjustment" and "economic recovery" programs. Austerity measures promoted by the World Bank and the IMF resulted in many governmental and parastatal properties being sold off to private concerns. Extension staffs and agricultural subsidies were drastically slashed, and many rural cropping systems were radically reconfigured in response. Under these circumstances, women were often encouraged to pursue independent economic interests that would have been much more circumscribed under other conditions (Thiesen et al. 1989). They were also negatively affected in the sense that the decline in government services forced them to intensify their work routines in order to secure their own and their families' livelihoods (Mackenzie 1993).

CHAPTER 2.   THE RISE OF A FEMALE CASH CROP

1. Weil indicates that the relationship between escalating food costs and intensified groundnut production extended well into this century: "between 1945 and 1970 there was a mass movement of other communities to areas near the swamps. Both men and women were attuned to the varying market price of rice, and they were acutely aware of [its] replacement costs. . . . Data from Kiang [a Lower River Division district on the South Bank] for the first half of the 1960s show that 20–30 percent of the cash derived from peanut sales was used directly to purchase food, mainly rice, and much of the remainder was used to pay debt service on loans taken out to purchase food. . . . The result was (and is) that the majority of peanut-based income was (and is) directly or indirectly used to purchase rice" (Weil 1986, 301).

2. Variations of this arrangement are common throughout much of West Africa (Funk 1988; Hemmings-Gapihan 1982; Longhurst 1982; MacCormack 1982; Roberts 1988).

3. Nagarajan, Meyer, and Graham's (1994) study of the export potential of The Gambia's horticultural enclaves is also directed primarily at the large capi-

talist and contract growers in Western Division. Compare Holcomb's (1991) assessment of the relative "[in]visibility" of North Bank production.

4. With qualitative data from only a single year, it is impossible to draw anything but the broadest conclusions regarding crop system changes. This caveat notwithstanding, the general profile drawn up by Gamble (1955) is a reasonable approximation of the region's agricultural practices during that period.

5. The tradeoff in producing short-duration varieties is that they are typically less productive than those with longer growing seasons. Many hybrid seed varieties are also less resistant to pests and disease than locally developed varieties.

6. These data were provided by the World Bank's WID project in The Gambia in 1991.

7. As Cooper (1987) notes, however, the multiple objectives of garden projects are often incompatible. In particular, the caloric expenditure generated through the hard labor of hand irrigation often offsets the nutritional value of vitamin-rich vegetable crops (see also Schoonmaker-Freudenberger 1991).

8. As Jabara (1990) notes, these figures must be treated with caution given that they reflect extensive trans-shipments of rice from the Gambian port to other nations in the West African interior.

9. These changes are expressed in nominal terms, much of the variation being due to depreciation of the *dalasi* currency in 1986. Such perturbations in the national accounts were seen by rice traders as an invitation to windfall profits, however, and unofficial prices in most areas ran much higher. See, for example, the anecdote in Hudson (1991, 212f).

10. These are approximate dates and locations; the information was provided by Public Works Department and Area Council employees in Kerewan.

11. To cite but two brief examples of the far-reaching impact of this commodification, fathers of newborn children were expected to provide several containers of powdered infant formula to their wives as part of their children's naming ceremonies. Similarly, in the marriage process, which typically involves a series of exchanges between husbands and wives and / or their families, husbands were expected to buy their wives expensive imported cosmetics. These and other imported commodities became more readily available as a result of the growth of the *lumoo* border markets.

12. Nominal prices rose spectacularly in response to currency devaluation. For details, see Johm (1990, 87).

13. Johm (1990) reports declines in fertilizer use in excess of 50 percent for groundnut production; my own survey of seventy-six men in Kerewan indicated that not a single male farmer in the sample had purchased fertilizer at 1991 prices. Several informants invoked the same rationale for not buying fertilizer under the conditions imposed by structural adjustment: "A bag of fertilizer now costs as much as a bag of rice. Which would you choose to buy?" It is also worth noting in passing that the decline in fertilizer use by male farmers stands in sharp contrast with the almost universal (98%) application of chemical fertilizer by women vegetable growers in Kerewan at the same time.

CHAPTER 3.   GONE TO THEIR SECOND HUSBANDS

1. For reviews of this literature, see Dwyer and Bruce 1988; Guyer and Peters 1987; and Stichter and Parpart 1988. Recent monographs addressing this topic include Clark 1994; Goheen 1996; and Leach 1994.

2. Husbands and wives commonly occupy separate living quarters. Women sleep with their husbands on a rotational basis with other co-wives; a rotation normally lasts for two days and nights, during which time the wife "on duty" also cooks and cleans for her husband. It is a traditional sign of deference for a woman to go to her husband's sleeping quarters first thing in the morning and greet him with a curtsey *(saama)* before going about her daily affairs. This is especially the case when a young woman is married to a much older man. Women interviewed on this topic admitted that garden work sometimes interfered with this practice: "Yes, it's true because a man may go [to the mosque] for dawn prayers and continue on some errands in the village. Before he comes back home, the wife may leave for the garden without seeing him." In most cases, however, women went out of their way to continue performing this highly symbolic gesture (see further discussion in chapter 4).

3. Data from Niumi Lameng in 1989 showed even higher average incomes (D2190), but this was not a random sample.

4. The Gambian *dalasi* was exchanged at a rate of D7.35=$1.00 in 1991, when these data were collected.

5. By contrast, proximity to Banjul allowed growers in Niumi Lameng to sell up to 90 percent of their produce to traveling vegetable buyers at the garden gate; they thus escaped the onerous burden of transportation costs experienced in Kerewan.

6. These results should be interpreted with caution. First, the comparison necessarily selected for households where men were groundnut farmers and omitted others. And second, the available data for men's and women's incomes were not drawn from the same years. The comparison is based on the earnings of thirty-six women from the March–April 1989 Niumi Lameng sample, seventy-five women from the February–June 1991 Kerewan sample, and their respective husbands' earnings from the June–October 1990 groundnut season. The figures do generally support the contention that women earned much more cash than their husbands in many households, a finding that was amply supported by anecdotal evidence.

A somewhat different perspective is provided by Jabara et al. (1991). In a study that reports aggregate data for three North Bank villages, not all of which had large market gardens, estimates of per capita rural incomes from all sources (including imputed values for "own-produced food") show that vegetable income averaged 12 percent of total household income. This compared with 21 percent from groundnuts, 17 percent from cereals, 15 percent from gifts/remittances, and 14 percent from "business." At first glance, these figures appear to partially contradict the Kerewan-Niumi Lameng data. It is important to note, however, that the two surveys are based on different methodological premises. The aggregation of income data by Jabara et al. (1991) allows the incomes of a handful of wealthy businessmen and groundnut growers to skew the overall results. The survey thus downplays the relative importance of vegetable incomes at the individual household level. Moreover, the inclusion of own-produced food into the income equation slights the contribution of the horticultural sector further. Own-produced rice and millet are rarely if ever sold on the North Bank—they do not effectively enter into the cash economy, and control over their production bestows a different kind of power and prestige than the cash-in-hand transactions that characterize vegetable sales.

7. This marked a substantial contribution. Prices for rice ranged in the neighborhood of D200 per bag on the North Bank in 1991, a cost roughly equivalent to 13 percent of the annual per capita income of rural Gambians (Jabara et al. 1991).

8. In an informal survey conducted in 1989 of mostly well-to-do gardeners, eight of thirty-five women surveyed had purchased their family's ram or goat for the annual Islamic feast day of *Taboski (Id ul Kabir)*.

9. Although only 27 percent of the research sample reported paying for their children's schooling, not all women had children of school-going age; thus, the proportion of women with children in school who paid schooling expenses themselves was actually considerably higher.

10. For example, some men were instrumental in helping their wives negotiate access to land.

11. The loan of a donkey or ox cart was not an insignificant gesture. Since men controlled virtually all draught animals and farming equipment in The Gambia, vegetable growers would otherwise have been forced to carry hundreds of kilos of produce by headpan a kilometer or more to the village. Many women without the benefit of *diya* did so anyway.

12. A woman's response to her husband taking a second wife would rest very heavily on how both her husband and the new wife treated her once the new marriage agreement was effected. As the senior wife, she would expect to be accorded greater respect than her junior partner. The fact remains, however, that the presence of a co-wife was often critical to a woman's successful management of home and garden based work routines, as chapter 4 demonstrates.

CHAPTER 4.   BETTER HOMES AND GARDENS

1. The market for capsicum peppers, a late-season crop, followed suit a few weeks later.

2. This is the classic approach taken by agricultural geographers such as Von Thünen (Hall 1966) and Chisholm (1962).

3. The limits of this type of economic analysis should, however, be explicitly acknowledged. Women who were otherwise mainly confined to their villages derived significant *benefits* from travel to markets. Moreover, the ability to convert cash income into more secure assets such as jewelry, cloth, or livestock was also an extremely important option for women concerned with protecting their garden incomes from their husbands (see chap. 3).

4. Most traders in the *lumoolu* frequented by the Mandinka women in Kerewan were Wolof speakers. By virtue of their relative isolation in their home villages, the women were far less likely to learn second and third languages than

men and were therefore at a disadvantage when it came to negotiating trade agreements.

5. The same group dug 93 new wells and deepened or repaired 130 existing wells in 1990. The increase in cost from 1990 to 1991 is partially explained by the opening of new garden perimeters in 1991.

6. Women in one North Bank community in 1987 were forced to forgo the gardening season altogether because the town's well digger opted to rebuild his house during the dry season rather than hire his labor out to gardeners.

7. The proclamation prohibiting free-ranging livestock (Mandinka: *tongo*) is announced each year on a town-by-town basis as the rains begin.

8. Cattle were the exception to this rule. Although they were not as closely monitored as during the rains, local herds were often tethered at night on a rotational basis in farmers' fields as a way of building up soil fertility. Farmers worked out different forms of payment or exchange to pay the herd owners for the dung the cattle deposited, a critical factor in maintaining continuous cultivation of upland fields surrounding the community. The degree of ground cover left for cattle and other livestock to browse depended heavily on the extent of seasonal bush fires.

9. Gaye, Jack, and Caldwell (1986) report mean monthly air temperatures as high as 41° C. Baldeh (1993) cites a mean air temperature of 38° C. and an average pan evaporation of 6–8 mm / day during the rainy season. Dry season estimates are not available, but Baldeh concludes that "mean evaporation . . . exceeds available moisture from precipitation" almost year round (Baldeh 1993, 28).

10. This is especially true for "hard work" tasks (table 9).

11. The 19 hectares included in our detailed physical inventory of garden sites were expanded to 22.5 hectares in 1991. This explains the discrepancy between the physical survey data and the data in table 14.

12. For example, many women (57% of the research sample) lit smudge pots in the evening to ward off insects that become active at that time of day and feed in the cooler nighttime temperatures.

13. The presence of trees and ground cover made the gardens several degrees cooler on average than temperatures prevailing in town.

CHAPTER 5.   BRANCHING INTO OLD TERRITORY

1. A small percentage of land in swamps was originally cleared by men on behalf of their families. These plots were accordingly transmitted along male lines of inheritance. Also, in cases where a woman had no female heirs, her rice plots sometimes passed to her son, who would then reallocate them to his wives or daughters.

2. There were three lineages in my principal research site. The town chief was generally drawn from the first, and the town religious leader *(imam)* from the second. The third was the remnant of a warrior clan.

3. These women held prior *kono banko* rights to plots which were recognized after gardens were established in rice-growing areas.

4. These payments have ranged from D5.00 at the time the earliest gardens

were established in the mid-1970s up to D30.00 in 1995—the rough equivalent of $1–3—for plots averaging roughly 100 square meters in size.

5. Interestingly, none of my informants was aware of any association of the term with land allotments prior to the garden boom.

6. Women charged that this diversion of funds by landholders was also common in the case of subscriptions the garden groups impose on themselves for fence repair and well maintenance.

7. The derivation of the term *rango*, which was used somewhat idiomatically in this community to refer to individual plots, is the English term "rank," as in rank and file. Gardeners have applied it to plots in recognition of the geometric grid pattern superimposed on gardens by extension agents who frequently assisted gardeners with initial site surveys and subsequent allocation of plots.

8. Interestingly, *new* members of garden groups at the time of expansion were required to pay *kumakaalu*. The extension of rights for existing members was thus in some sense a means of claiming the value they themselves had added to the garden perimeters.

9. All names employed in this account are pseudonyms; Sanyang's identity is, in any case, well known throughout the area, and is a matter of public record. Certain minor details in the narrative have been fabricated to protect the identities of other individuals whose stories are told.

10. These fees amounted to less than two dollars, a sum small enough that most women would have little difficulty paying for a plot out of proceeds from sales of poultry or vegetables grown in "kitchen" gardens, but significant enough when multiplied by the number of growers to represent a sizable rent for the landholder.

11. The detailed citation of these records is omitted to protect the identity of individuals involved in the case.

12. Half of these women worked plots within Al-Haji Sanyang's perimeter; others were located in a second large perimeter, the "women's garden" alluded to above, which had subsequently been established on the south side of the village.

13. The exact timing of this intervention remains somewhat unclear. The lack of mention of trees in any of the early documentation on the controversy suggests that the issue arose rather late in the day.

14. The nursery attendant for the Forestry Department, who received several requests for technical assistance in relocating the trees, reported that many of the trees were too mature to be transplanted successfully and most of those that were young enough to move were transferred too quickly under time pressure from the commissioner's injunction.

15. This practice is not uncommon elsewhere (Fortmann and Bruce 1988; Raintree 1987).

16. Gamble's anthropological account of tree tenure principles operating in the area in the 1940s confirms this claim: "So far the problems arising from trees of commercial value, e.g., fruit trees, seems not to have arisen. The people of [a nearby community] say that a man may not plant fruit trees on another man's land without his permission. If the landowner objects he can have the trees torn up. If he gives permission it amounts to a permanent alienation of the land for

the granters would never claim it back. Others maintain that the original owners can claim the land, but that the trees remain the property of the planter." Gamble (1947), NAG, 9/399, 22–23. Witness also the judgment of a Muslim cleric on the South Bank of the river who described a typical dispute over tree tenure: "It has happened that when a man who has planted a tree subsequently dies, another person (possibly the owner of the land) can assume the responsibility of watering and caring for the tree. If, after some time passes, the son of the man who planted the tree claims ownership of the tree, a dispute may arise in which the caretaker claims ownership over the tree. The *imam* stated that resolution of such disputes is clear-cut—the act of watering and caring for the tree does not confer rights of ownership over the tree. The son of the tree planter inherits ownership rights to the tree" (Freudenberger and Sheehan 1994, 67).

17. Returns per hectare favored vegetables, while returns to labor favored tree planting. For gardeners with limited land resources, the former was decisive.

18. Other factors relevant to the Kerewan case are not accounted for in this comparison. Many of the trees planted in Kerewan's gardens were still quite immature at the time market gardens started to pay off; in some perimeters pest damage to fruit also diminished the productivity of tree crops; finally, plans to sell fruit were routinely undercut by the implicit demand from family members that harvests from the trees be shared with them before produce was sold on the open market (cf. Osborn 1989). The decisions taken to abandon fruit production in favor of gardening must be seen in this light.

19. While these different gender identities inflected by the garden boom were overlapping, they were not identical. There were only a few dozen women in Kerewan who did not garden by the early 1990s. Most of these women either were married to civil servants posted to Kerewan or held professional positions themselves. Very few were Kerewan natives. While these women did not join gardeners in their efforts to adjust to the shifting development agenda, they clearly participated in the complex negotiations over the nature of the conjugal contract in Mandinka society. By contrast, there were at most two dozen men who controlled the low-lying land parcels most coveted by gardeners. They were granted these rights by virtue of their relative seniority in their respective family lineages. Many of these men were themselves married to gardeners and shared the concerns of their male peers regarding household budgetary issues on that basis. But their interests in land management sometimes superseded these concerns. Thus the landholders found themselves at odds with other men in the community who had come to appreciate the benefits of the garden boom and aligned themselves with gardeners on that basis in their quest to expand their garden perimeters

CHAPTER 6.   CONTESTING
                AGROFORESTRY INTERVENTIONS

1. A large *Gmelina arborea* plantation in Lower River Division on The Gambia's South Bank was established according to similar principles under the colonial government in 1959. When the *Gmelina* seedlings were first planted, farm-

ers were granted privileges to plant groundnuts on the site between the rows of seedlings for one year. Land was free of charge, provided the farmers tended the tree seedlings. Subsequent efforts to establish teak forests under the same principles were frustrated, however, when farmers neglected and / or destroyed tree seedlings in order to prolong access to land for groundnut production (Thoma 1989).

2. Note the similarities between the women's attitudes toward woodlots and men's attitudes toward women's gardens described in the Preface. Neither group was willing to provide labor freely when their participation in projects was taken for granted and their share in the project benefits was not clearly specified.

3. The distinction between Kerewan and Swaray Kunda women is complicated by the fact that there is considerable intermarriage between families in the two communities.

4. This is a pseudonym.

5. The cutting of fence posts is a source of pressure on forest resources that developers have sought to minimize through the promotion of live fencing. The use of thorny planting materials in lieu of barbed wire is a practice that theoretically minimizes the cost of fence construction. It is intended to help reduce dependency on outside suppliers and thus renders fence maintenance more "sustainable" over the long term.

6. The Swaray site was planted with an exceptionally high tree density of ninety mangoes, fourteen oranges, and eighty bananas per hectare (Site 9, table 14). For comparison, see other Kerewan gardens in table 14.

7. These income estimates were not based on a random sample.

8. I am personally implicated in this particular outcome. The zoning practices adopted by my former assistant were the direct outcome of our research collaboration. Recommendations for similar approaches were part of my standard research briefing to development agencies engaged in horticulture planning during the principal research period four years earlier.

9. USAID itself closed down its Gambian mission in 1996. The coup served as a pretext for adding The Gambia to a long list of African mission closures by USAID in the early 1990s.

10. An earlier version of the argument in this section appeared in Schroeder and Suryanata (1996).

11. This approach was subsequently promoted in many other parts of the world, including The Gambia. See chapter 6, n. 1, above; cf. Goswami (1988); King (1988); and Peluso (1992).

Works Cited

Adams, Jonathan, and Thomas McShane. 1992. *The myth of wild Africa: Conservation without illusion.* New York: Norton.

Agarwal, Bina. 1992. The gender and environment debate: Lessons from India. *Feminist Studies* 18 (1):119–158.

Agroprogress International. 1990. Project preparation consultancy for an integrated development programme [European Community] for the North Bank Division. Bonn: Agroprogress International.

Altieri, Miguel, ed. 1987. *Agroecology: The scientific basis of alternative agriculture.* Boulder: Westview Press.

Anderson, Anthony, ed. 1990. *Alternatives to deforestation: Steps toward sustainable use of the Amazon rain forest.* New York: Columbia University Press.

Anderson, David, and Richard Grove, eds. 1987. *Conservation in Africa: People, policies and practice.* New York: Cambridge University Press.

Arias, Margarita. 1992. Mothering the earth. *Development* 1992 (1):75–78.

Baldeh, Falie. 1993. Environmental profile of AATG's rural development areas. *Yiriwa Kibaro* 1 (1). Banjul, The Gambia: Action Aid.

Bandarage, Asoka. 1984. Women in development: Liberalism, Marxism and Marxist feminism. *Development and Change* 15 (4):495–515.

Barker, Jonathan. 1989. *Rural communities under stress: Peasant farmers and the state in Africa.* Cambridge: Cambridge University Press.

Barrett, Hazel, and Angela Browne. 1989. Time for development? The case of women's horticultural schemes in rural Gambia. *Scottish Geographical Magazine* 105 (1):4–11.

———. 1991. Environmental and economic sustainability: Women's horticultural production in The Gambia. *Geography* 76:241–248.

Bassett, Thomas. 1993. Introduction: The land question and agricultural transformation in Sub-Saharan Africa. In *Land in African agrarian systems,* ed.

Thomas Bassett and Donald Crummey, pp. 3–34. Madison: University of Wisconsin Press.

Beneria, Lourdes, and Gita Sen. 1981. Accumulation, reproduction, and women's role in economic development: Boserup revisited. *Signs* 7 (2):279–298.

Bernstein, Henry. 1982. Notes on capital and peasantry. In *Rural development: Theories of peasant economy and agrarian change*, ed. John Harriss, pp. 160–177. London: Hutchinson University Library.

Berry, Sara. 1984. The food crisis and agrarian change in Africa. *African Studies Review* 27 (2):59–112.

———. 1987. Macro-policy implications of research on rural households and farming systems. In *Understanding Africa's rural households and farming systems*, ed. Joyce Moock, pp. 199–216. Boulder: Westview Press.

———. 1989. Social institutions and access to resources. *Africa* 59 (1):41–55.

———. 1993. *No condition is permanent: The social dynamics of agrarian change in sub-Saharan Africa*. Madison: University of Wisconsin Press.

Biodiversity Support Program (BSP). 1993. *African biodiversity: Foundation for the future*. Washington, DC: World Wildlife Fund, Nature Conservancy, and World Resources Institute, with the U.S. Agency for International Development.

Blaikie, Piers. 1985. *The political economy of soil erosion in developing countries*. London: Longman.

Blaikie, Piers, and Harold Brookfield. 1987. *Land degradation and society*. London: Methuen.

Bonner, Raymond. 1993. *At the hand of man: Peril and hope for Africa's wildlife*. New York: Vintage Books.

Boserup, Ester. 1970. *Women's role in economic development*. London: Allen & Unwin.

Boughton, Duncan, and Jacqueline Novogratz. 1989. *The Gambia: Women in development project, agricultural component*. Government of the Gambia / World Bank.

Braidotti, Rosi, Ewa Charkiewicz, Sabine Häusler, and Saskia Wieringa, eds. 1994. *Women, the environment and sustainable development: Towards a theoretical synthesis*. London: Zed Press.

Brown, Michael, and Wyckoff-Baird, Barbara. 1992. *Designing integrated conservation and development projects*. Washington, DC: Biodiversity Support Program.

Bruce, John. 1993. Do indigenous tenure systems constrain agricultural development? In *Land in African agrarian systems*, ed. Thomas Bassett and Donald Crummey, pp. 35–37. Madison: University of Wisconsin Press.

Bruce, Judith, and Daisy Dwyer. 1988. Introduction. In *A home divided: Women and income in the third world*, ed. Daisy Dwyer and Judith Bruce, pp. 1–19. Stanford: Stanford University Press.

Brush, Stephen, and Doreen Stabinsky, eds. 1996. *Valuing local knowledge: Indigenous people and intellectual property rights*. Washington, DC: Island Press.

Bryant, Raymond. 1992. Political ecology: An emerging research agenda in Third World studies. *Political Geography* 11 (1):12–36.

———. 1994. The rise and fall of *taungya* forestry: Social forestry in defence of the Empire. *Ecologist* 24 (1):21–26.

———. 1997. Beyond the impasse: The power of political ecology in Third World environmental research. *Area* 29 (1):5–19.

Bryant, Raymond, and Sinead Bailey. 1997. *Third World political ecology*. London: Routledge.

Campbell, Gerald, and Lisa Daniels. 1987. Horticultural marketing mission to The Gambia. Gambian Agricultural Research and Diversification Report No. 16. Banjul: Ministry of Agriculture / Department of Agricultural Research.

Carney, Judith. 1986. The social history of Gambian rice production: An analysis of food security strategies. Ph.D. diss. University of California–Berkeley.

———. 1988a. Struggles over crop rights and labour within contract farming households in a Gambian irrigated rice project. *Journal of Peasant Studies* 15 (3):334–349.

———. 1988b. Struggles over land and crops in an irrigated rice scheme: The Gambia. In *Agriculture, women and land*, ed. Jean Davison, pp. 59–78. Boulder: Westview Press.

———. 1992. Peasant women and economic transformation in The Gambia. *Development and Change* 23 (2):67–90.

———. 1993. Converting the wetlands, engendering the environment: The intersection of gender with agrarian change in The Gambia. *Economic Geography* 69 (3):329–348.

Carney, Judith, and Michael Watts. 1990. Manufacturing dissent: Work, gender and the politics of meaning in a peasant society. *Africa* 60 (2):207–241.

———. 1991. Disciplining women? Rice, mechanization and the evolution of gender relations in Senegambia. *Signs* 16 (4):651–681.

Chisholm, Michael. 1962. *Rural settlement and land use: An essay in location*. London: Hutchinson.

Chowdhury, Geeta. 1995. Engendering development? Women in development (WID) in international development regimes. In *Feminism postmodernism development*, eds. Marianne Marchand and Jane Parpart, pp. 26–41. London: Routledge.

Clark, Gracia. 1994. *Onions are my husband*. Chicago: University of Chicago Press.

Clay, Jason. 1988. *Indigenous peoples and tropical forests: Models of land use and management from Latin America*. Cambridge, MA: Cultural Survival.

Cleaver, Kevin, and Götz Schreiber. 1993. *The population, agriculture and environment nexus in sub-Saharan Africa*. Washington, DC: World Bank.

Cline-Cole, Reginald. 1997. Promoting (anti-)social forestry in northern Nigeria? *Review of African Political Economy* 74:515–536.

Cooper, Barbara. 1987. Survey of the literature on gardens and gardening projects with emphasis on West Africa. Unpublished paper. Boston: Oxfam America.

Daniels, Lisa. 1988. The economics of staggered production and storage for selected horticultural crops in The Gambia. M.S. Thesis, University of Wisconsin.

Davison, Jean, ed. 1988. *Agriculture, women and land.* Boulder: Westview Press.

de Janvry, Alain. 1981. *The agrarian question and reformism in Latin America.* 2nd ed. Baltimore: Johns Hopkins University Press, 1981.

de Oliveira, R. D., and T. Corral, eds. 1992. *Terra femina.* Rio de Janeiro: IDAC / REDEH.

DeCosse, Philip, and E. Camara. 1990. A profile of the horticultural production sector in The Gambia. Paper prepared for the National Horticultural Policy Workshop, Banjul, The Gambia.

Deere, Carmen Diana, and Magdalena León, eds. 1987. *Rural women and state policy: Feminist perspectives on Latin American agricultural development.* Boulder: Westview Press.

Department of Planning (Government of The Gambia). 1991. National agricultural statistical survey, 1990. Banjul: Ministry of Agriculture.

Dey, Jennie. 1981. Gambian women: Unequal partners in rice development projects? *Journal of Development Studies* 17 (3):109–122.

———. 1982. Development planning in The Gambia: The gap between planners' and farmers' perceptions, expectations and objectives. *World Development* 10 (5):377–396.

———. N.d. The socio-economic organization of farming in The Gambia and its relevance for agricultural development planning. Agricultural Administration Network Papers No. 7. London: Overseas Development Institute.

Dove, Michael. 1990. Socio-political aspects of home gardens in Java. *Journal of Southeast Asian Studies* 21 (1):155–163.

Dwyer, Daisy, and Judith Bruce, eds. 1988. *A home divided: Women and income in the third world.* Stanford: Stanford University Press.

The Ecologist. 1993. *Whose common future? Reclaiming the commons.* Philadelphia: New Society.

Escobar, Arturo. 1995. *Encountering development: The making and unmaking of the Third World.* Princeton: Princeton University Press.

Farrell, John. 1987. Agroforestry systems. In *Agroecology: The scientific basis of alternative agriculture,* ed. M. Altieri, pp. 149–158. Boulder: Westview Press.

Ferguson, James. 1990. *The anti-politics machine: "Development," depoliticization and bureaucratic power in Lesotho.* Cambridge: Cambridge University Press.

Folbre, Nancy. 1986. Hearts and spades: Paradigms of household economics. *World Development* 14 (2):245–255.

Foley, Glenn. 1994. The Gambia energy sector review: Woodfuels and household energy, new and renewal energy sources, rural electrification. Report prepared for the Nordic Consulting Group. Copenhagen, Oslo, and Stockholm.

Fortmann, Louise, and John Bruce, eds. 1988. *Whose trees? Proprietary dimensions of forestry.* Boulder: Westview Press.

Fortmann, Louise, and Dianne Rocheleau. 1985. Women and agroforestry: Four myths and three case studies. *Agroforestry Systems* 2:253–272.

Franke, Richard, and Barbara Chasin. 1980. *Seeds of famine: Ecological destruction and the development dilemma in the West African Sahel.* Totowa, NJ: Rowman and Allanheld.

Freudenberger, Mark. 1994. Tenure and natural resources in The Gambia: Summary of research findings and policy options. Madison: University of Wisconsin Land Tenure Center.

Freudenberger, Mark, and Nancy Sheehan. 1994. Tenure and resource management in The Gambia: A case study of the Kiang West District. Madison: University of Wisconsin Land Tenure Center.

Fried, Stephanie Gorson. Forthcoming. Tropical forests forever? A contextual ecology of Bentian Rattan agroforestry systems. In *People, plants and justice: Resource extraction and conservation in tropical developing countries,* ed. Charles Zerner. New York: Columbia University Press.

Funk, Ursula. 1988. Land tenure, agriculture, and gender in Guinea-Bissau. In *Agriculture, women and land,* ed. Jean Davison, pp. 33–58. Boulder: Westview Press.

Gamble, David. 1947. Kerewan—A Mandinka village: Social structure and daily life. NAG / 9 / 399, 22–23. National Archives of The Gambia, Banjul.

———. 1949. *Contributions to a socio-economic survey of The Gambia.* London: Colonial Office.

———. 1955. *Economic conditions in two Mandinka villages: Kerewan and Keneba.* London: Colonial Office.

Gaye, G. O., Isatou Jack, and John Caldwell. 1986. Use of farming systems research / extension (FSR / E) methods to identify research priorities in The Gambia, West Africa. Working Paper No. 1. Gambian Agricultural Research Program. Yundum, The Gambia: Ministry of Agriculture / Department of Agricultural Research.

Gladwin, Christina, and Della McMillan. 1989. Is a turnaround in Africa possible without helping African women to farm? *Economic Development and Cultural Change* 37:345–369.

Goheen, Miriam. 1996. *Men own the fields, women own the crops: Gender and power in the Cameroon grassfields.* Madison: University of Wisconsin Press.

Goodman, David, and Michael Redclift. 1991. *Re-fashioning nature: Food, ecology and culture.* London: Routledge.

Goswami, P. C. 1988. Agro-forestry: Practices and prospects as a combined land-use system. In *Whose trees? Proprietary dimensions of forestry,* eds. Louise Fortmann and John Bruce, pp. 297–300. Boulder: Westview Press.

GOTG (Government of The Gambia). 1992. *The Gambia Environmental Action Plan, 1992–2001.* Banjul, The Gambia: GOTG.

Gregory, James. 1984. The myth of the male ethnographer and the woman's world. *American Anthropologist* 86 (2): 316–327.

Guha, Ramachandra. 1989. *The unquiet woods: Ecological change and peasant resistance in the Himalaya.* New Delhi: Oxford University Press.

Guyer, Jane. 1980. Food, cocoa and the division of labor by sex. *Comparative Studies in Society and History* 22: 355–373.

———. 1984. Naturalism in models of African production. *Man* 19: 371–388.

————. 1988. Dynamic approaches to domestic budgeting: Cases and methods from Africa. In *A home divided*, eds. Daisy Dwyer and Judith Bruce, pp. 155–172. Stanford: Stanford University Press.

Guyer, Jane, and Pauline Peters, eds. 1987. Special issue: Conceptualizing the household: Theory and policy in Africa. *Development and Change* 18 (2).

Hall, Peter, ed. 1966. *Von Thünen's Isolated state (an English edition of Der Isolierte Staat, by Johann Heinrich von Thünen)*. Oxford: Pergamon Press.

Haraway, Donna. 1989. *Primate visions: Gender, race, and nature in the world of modern science*. New York: Routledge.

Harris, Olivia. 1981. Households as natural units. In *Of marriage and the market*, eds. Kate Young, Carol Wolkowitz, and Roslyn McCullagh, pp. 49–68. London: CSE Books.

Haswell, Margaret. 1975. *The nature of poverty: A case history of the first quarter-century after World War II*. New York: St. Martin's Press.

Hecht, Susanna, Anthony Anderson, and Peter May. 1988. The subsidy from nature: Shifting cultivation, successional palm forests, and rural development. *Human Organization* 47:25–35.

Hemmings-Gapihan, Grace. 1982. International development and the evolution of women's economic roles: A case study from Northern Gulma, Upper Volta. In *Women and work in Africa*, ed. Edna Bay, pp. 171–190. Boulder: Westview Press.

Hodgson, Dorothy, and Sheryl McCurdy. Forthcoming. *"Wicked" women and the reconfiguration of gender in Africa*. Portsmouth, NH: Heinemann Press.

Holcomb, Jan. 1991. An assessment of post-harvest and marketing practices of women vegetable producers impacted by the AFSI (African Food Systems Initiative) program in The Gambia. Banjul: U.S. Peace Corps / The Gambia, April.

Hudson, Mark. 1991. *Our grandmothers' drums*. New York: Holt.

IIED (International Institute for Environment and Development). 1994. *Whose Eden? An overview of community approaches to wildlife management*. London: IIED.

Jabara, Cathy. 1990. *Economic reform and poverty in The Gambia: A survey of pre- and post-ERP experience*. Monograph No. 8. Ithaca: Cornell Food and Nutrition Policy Program.

Jabara, Cathy, Marjatta Tolvanen, Mattias Lundberg, and Rohey Wadda. 1991. Incomes, nutrition, and poverty in The Gambia: Results from the CFNPP household survey. Washington, DC: Cornell Food and Nutrition Policy Program.

Jack, Isatou. 1990. Export constraints and potentialities for Gambian horticultural produce. Paper prepared for the National Horticultural Planning Workshop, Banjul, The Gambia.

Jackson, Cecile. 1993. Doing what comes naturally? Women and environment in development. *World Development* 21 (12):1947–1963.

————. 1995. From conjugal contracts to environmental relations: Some thoughts on labour and technology. *IDS Bulletin* 26 (1):33–39.

Jiggins, Janice. 1994. *Changing the boundaries: Women-centered perspectives on population and the environment*. Washington, DC: Island Press.

Johm, Ken. 1990. Policy issues and options for agricultural development in The Gambia. In *Structural adjustment, agriculture and nutrition: Policy options in The Gambia,* eds. Joachim von Braun, Ken Johm, Sambou Kinteh, and Detlev Puetz, pp. 84–91. Washington, DC: International Food Policy Research Institute (IFPRI).

Jones, Christine. 1986. Intra-household bargaining in response to the introduction of new crops: A case study from North Cameroon. In *Understanding Africa's rural households and farming systems,* ed. Joyce Moock, pp. 105–123. Boulder: Westview Press.

Kabeer, Naila. 1994. *Reversed realities: Gender hierarchies in development thought.* London: Verso.

Kandeh, H. B. S., and Paul Richards. 1996. Rural people as conservationists: Querying neo-Malthusian assumptions about biodiversity in Sierra Leone. *Africa* 66 (1):90–103.

Kebbede, Girma, and Mary J. Jacob. 1988. Drought, famine and the political economy of environmental degradation in Ethiopia. *Geography* 73 (1): 65–70.

King, K. F. S. 1988. Agri-silviculture (taungya system): The law and the system. In *Whose trees? Proprietary dimensions of forestry,* eds. Louise Fortmann and John Bruce, pp. 301–305. Boulder: Westview Press.

Kinteh, Sambou. 1990. A review of agricultural policy before and after adjustment. In *Structural adjustment, agriculture and nutrition: Policy options in The Gambia,* eds. Joachim von Braun, Ken Johm, Sambou Kinteh, and Detlev Puetz, pp. 5–26. Washington, DC: International Food Policy Research Institute (IFPRI).

Lawry, Steven. 1988. Report on Land Tenure Center mission to The Gambia. Madison: University of Wisconsin Land Tenure Center.

Leach, Melissa. 1991. Engendered environments: Understanding natural resource management in the West African forest zone. *IDS Bulletin* 22 (4): 17–24.

———. 1992. Women's crops in women's spaces: Gender relations in Mende rice farming. In *Bush base: Forest farm: Culture, environment and development,* eds. Elisabeth Croll and David Parkin, pp. 76–96. London: Routledge Press.

———. 1994. *Rainforest relations: Gender and resource use among the Mende of Gola, Sierra Leone.* Washington, DC: Smithsonian Institution Press.

Leach, Melissa, Susan Joekes, and Cathy Green. 1995. Editorial: Gender relations and environmental change. *IDS Bulletin* 26 (1): 1–8.

Leach, Melissa, and Robin Mearns, eds. 1996. *The lie of the land: Challenging received wisdom on the African environment.* Portsmouth, NH: Heinemann.

Longhurst, Richard. 1982. Resource allocation and the sexual division of labor: A case study of a Moslem Hausa village. In *Women and development: The sexual division of labor in rural economies,* ed. Lourdes Beneria, pp. 95–118. Geneva: ILO.

MacCormack, Carolyn. 1982. Control of land, labor and capital in rural southern Sierra Leone. In *Women and work in Africa,* ed. Edna Bay, pp. 35–54. Boulder: Westview Press.

Mackenzie, Fiona. 1990. Gender and land rights in Murang'a District, Kenya. *Journal of Peasant Studies* 17 (4):609–643.

———. 1991. Political economy of the environment, gender and resistance under colonialism: Murang'a District, Kenya, 1910–1950. *Canadian Journal of African Studies* 25 (2):226–256.

———. 1993. Exploring the contradictions: Structural adjustment, gender and the environment. *Geoforum* 24 (1):71–87.

———. 1994. "A piece of land never shrinks": Reconceptualizing land tenure in a smallholding district, Kenya. In *Land in African agrarian systems*, eds. Thomas Bassett and Donald Crummey, pp. 194–221. Madison: University of Wisconsin Press.

Mackintosh, Maureen. 1989. *Gender, class and rural transition: Agribusiness and the food crisis in Senegal*. London: Zed Press.

MacRae, Eric. 1992. Drought in Zimbabwe: The dynamics of disaster. *Geographical Magazine* 64 (9):12.

Madge, Clare. 1994. Collected food and domestic knowledge in The Gambia, West Africa. *Geographical Journal* 160 (3):280–294.

———. 1995a. The adaptive performance of West African life: Continuity and change of collecting activities in The Gambia. *Geografiska Annaler* 77B (2):109–124.

———. 1995b. Ethnography and agroforestry research: A case study from The Gambia. *Agroforestry Systems* 32:127–146.

Mann, Robert. 1990. Time running out: The urgent need for tree planting in Africa. *Ecologist* 20 (2): 48–53.

Marchand, Marianne, and Jane Parpart, eds. 1995. *Feminism postmodernism development*. London: Routledge.

Mbilinyi, Marjorie. 1991. *Big slavery: Agribusiness and the crisis in women's employment in Tanzania*. Dar es Salaam: Dar es Salaam University Press.

McPherson, Malcolm, and Steven Radelet, eds. 1995. *Economic recovery in The Gambia: Insights for adjustment in sub-Saharan Africa*. Cambridge: Harvard Institute for International Development.

Meldrum, Andrew. 1992. The big scorcher. *Africa Report* 37 (3):25–27.

Merchant, Carolyn. 1992. *Radical ecology: The search for a livable world*. New York: Routledge.

Michon, Genevieve, Hector de Foresta, Kusworo, and Philip Levang. Forthcoming. Formal recognition of farmer's rights as a pre-condition for the rebuilding of productive and durable community forests in Indonesia: The Damar agroforests in Krui, Sumatra. In *People, plants and justice: Resource extraction and conservation in tropical developing countries*, ed. Charles Zerner. New York: Columbia University Press.

Mickelwait, D., M. Riegelman, and C. Sweet. 1976. *Women in rural development*. Boulder: Westview Press.

Migot-Adholla, S., and Bruce John. 1993. Introduction: Are indigenous African tenure systems insecure? In *Searching for land tenure security in Africa*, eds. John Bruce and S. Migot-Adholla, pp. 1–14. Dubuque, IA: Kendall / Hunt.

Mikell, Gwendolyn. 1997. *African feminism: The politics of survival in sub-Saharan Africa*. Philadelphia: University of Pennsylvania Press.

Momsen, Janet. 1991. *Women and development in the Third World*. London: Routledge.

Moock, Joyce Lewinger. 1986. *Understanding Africa's rural households and farming systems*. Boulder: Westview Press.

Moore, Henrietta, and Megan Vaughan. 1994. *Cutting down trees: Gender, nutrition, and agricultural change in the Northern Province of Zambia, 1890–1990*. Portsmouth, NH: Heinemann.

Muntemba, Maud. 1982. Women and agricultural change in the railway region of Zambia: Dispossession and counterstrategies, 1930–1970. In *Women and work in Africa*, ed. Edna Bay, pp. 83–104. Boulder: Westview Press.

Nagarajan, Geetha, Richard Meyer, and Douglas Graham. 1994. Exporting fresh fruits and vegetables to Europe: Potential and constraints for Gambian producers. Washington, DC: USAID.

Nair, P. K. R. 1989. *Agroforestry systems in the tropics*. Boston: Kluwer Academic and ICRAF.

———. 1990. *The prospects for agroforestry in the tropics*. World Bank Technical Paper No. 131. Washington, DC: World Bank.

Nath, Kamla. 1985a. Labor-saving techniques in food processing: Rural women and technological change in The Gambia. African Studies Working Papers No. 108. Boston: Boston University.

———. 1985b. Women and vegetable gardens in The Gambia: Action Aid and rural development. African Studies Working Papers No. 109. Boston: Boston University.

Neumann, Roderick. 1995. Ways of seeing Africa: Colonial recasting of African society and landscape in Serengeti National Park. *Ecumene* 2: 149–69.

———. 1996. Dukes, earls and ersatz Edens: Aristocratic nature preservationists in colonial Africa. *Environment and Planning. D, Society & Space* 14 (1): 79–98.

Nicholson, Sharon. 1985. Sub-Saharan rainfall 1981–84. *Journal of Climate and Applied Meteorology* 24:1388–1391.

———. 1993. An overview of African rainfall fluctuations of the last decade. *Journal of Climate* 6:1463–1466.

Norton, G., Bradford Mills, Elon Gilbert, M. S. Sompo-Ceesay, and John Rowe. 1989. Analysis of agricultural research priorities in The Gambia. Unpublished manuscript.

Norton-Staal, Sarah. 1991. Women and their role in the agriculture and natural resource sector in The Gambia. Banjul: U.S. Agency for International Development.

Ortner, Sherry. 1974. Is female to male as nature is to culture? In *Women, culture, & society*, eds. Michelle Rosaldo and Louise Lamphere, pp. 67–87. Stanford: Stanford University Press.

Osborn, Elisabeth. 1989. Tree tenure: The distribution of rights and responsibilities in two Mandinka villages. M.S. Thesis, University of California–Berkeley.

Palmer, Ingrid. 1991. *Gender and population in the adjustment of African economies: Planning for change*. Geneva: ILO.

Parpart, Jane. 1989. *Women and development in Africa*. Lanham, MD: University Press of America.

————. 1995. Deconstructing the development "expert": Gender, development and the "vulnerable groups." In *Feminism postmodernism development*, eds. Marianne Marchand and Jane Parpart, pp. 221–243. London: Routledge.

Parpart, Jane, and Marianne Marchand. 1995. Exploding the canon: An introduction / conclusion. In *Feminism postmodernism development*, eds. Marianne Marchand and Jane Parpart, pp. 1–22. London: Routledge.

Peet, Richard, and Michael Watts. 1996. Liberation ecology: Development, sustainability, and environment in an age of market triumphalism. In *Liberation ecologies: Environment, development, social movements*, eds. Richard Peet and Michael Watts, pp. 1–45. London: Routledge.

Peluso, Nancy. 1992. *Rich forests, poor people: Resource control and resistance in Java*. Berkeley: University of California Press.

————. 1996. Fruit trees and family trees in an anthropogenic forest: Ethics of access, property zones, and environment change in Indonesia. *Comparative Studies in Society and History* 38 (3): 510–548.

Plumwood, Val. 1993. *Feminism and the mastery of nature*. New York: Routledge.

Posner, Joshua, and Elon Gilbert. 1987. District agricultural profile of Central Baddibu, North Bank Division. Gambian Agricultural Research Papers No. 2. Banjul: Department of Agricultural Research.

Pred, Allan. 1990. In other wor(l)ds: Fragmented and integrated observations on gendered languages, gendered spaces and local transformation. *Antipode* 22: 33–52.

Puetz, Detlev, and Joachim von Braun. 1990. Price policy under structural adjustment: Constraints and effects. In *Structural adjustment, agriculture and nutrition: Policy options in The Gambia*, eds. Joachim von Braun, Ken Johm, Sambou Kinteh, and Detlev Puetz, pp. 40–54. Washington, DC: International Food Policy Research Institute (IFPRI).

Raintree, John, ed. 1987. *Land, trees and tenure*. Nairobi, Kenya, and Madison: ICRAF and University of Wisconsin Land Tenure Center.

Rangan, Haripriya. 1993. Romancing the environment: Popular environmental action in the Garhwal Himalayas. In *In defense of livelihood: Comparative studies on environmental action*, eds. John Friedmann and Haripriya Rangan, pp. 155–181. West Hartford, CT: Kumarian Press.

————. 1996. From Chipko to Uttaranchal: Development, environment, and social protest in the Garhwal Himalayas, India. In *Liberation ecologies: Environment, development, social movements*, eds. Richard Peet and Michael Watts, pp. 205–226. London: Routledge.

Rathgeber, Eva. 1995. Gender and development in action. In *Feminism postmodernism development*, eds. Marianne Marchand and Jane Parpart, pp. 204–220. London: Routledge.

Redford, Kent, and Christine Padoch, eds. 1992. *Conservation of neotropical forests: Working from traditional resource use*. New York: Columbia University Press.

Ribot, Jesse. 1995. From exclusion to participation: Turning Senegal's forest policy around? *World Development* 23 (9): 1587–1599.

Ridder, R. M. 1991. Land use inventory for The Gambia on the basis of Landsat-TM scenes including a comparison with previous investigations. Feldkirchen, Germany: GTZ Deutsche Fortservice.

Robbins, David. 1983. Drought plagues southern Africa. *Africa Report* 28 (4):56–58.

Roberts, Penelope. 1979. "The integration of women into the development process": Some conceptual problems. *IDS Bulletin* 10 (3):60–66.

———. 1988. Rural women's access to labor in West Africa. In *Patriarchy and class*, eds. Sharon Stichter and Jane Parpart, pp. 97–114. Boulder: Westview Press.

Roberts, Susan B., A. A. Paul, T. J. Cole, and R. G. Whitehead. 1982. Seasonal changes in activity, birth-weight and lactational performance in rural Gambian women. *Transactions of the Royal Society of Tropical Medicine and Hygiene*, 76:668–678.

Rocheleau, Dianne. 1987. Women, trees and tenure: Implications for agroforestry research and development. In *Land, trees and tenure*, ed. John Raintree, pp. 79–121. Nairobi, Kenya, and Madison: ICRAF and University of Wisconsin Land Tenure Center.

———. 1988. Gender, resource management and the rural landscape: Implications for agroforestry and farming systems research. In *Gender issues in farming systems research and extension*, eds. S. Poats, M. Schmink, and A. Spring. Boulder: Westview Press.

———. 1991. Gender, ecology and the science of survival: Stories and lessons from Kenya. *Agriculture and Human Values* 8 (1 / 2):156–165.

———. 1995. Gender and biodiversity: A feminist political ecology perspective. *IDS Bulletin* 26 (1): 9–16.

Rocheleau, Dianne, and Linda Ross. 1995. Trees as tools, trees as text: Struggles over resources in Zambrana Chacuey, Dominican Republic. *Antipode* 27(4): 407–428.

Rocheleau, Dianne, Barbara Thomas-Slayter, and David Edmunds. 1995. Gendered resource mapping: Focusing on women's spaces in the landscape. *Cultural Survival Quarterly* (Winter):62–68.

Rocheleau, Dianne, Barbara Thomas-Slayter, and Esther Wangari. 1996. *Feminist political ecology: Global issues and local experiences*. New York: Routledge.

Rodda, Annabel, ed. 1991. *Women and the environment*. London: Zed Books.

Roe, Emery. 1998. *Except—Africa: Remaking development, rethinking power*. New Brunswick, NJ: Transaction.

Sachs, Wolfgang, ed. 1992. *The development dictionary*. London: Zed Books.

———. 1993. *Global ecology: A new arena of political conflict*. London: Zed Books.

Schindele, Wolfgang, and Foday Bojang. 1995. Gambian forest management concept, part I: the forest sector of The Gambia. GGFP Report No. 29. Feldkirchen, Germany: GTZ / Deutsche Forstservice

Schoepf, Brooke, and Claude Schoepf. 1988. Land, gender and food security in Eastern Kivu, Zaire. In *Agriculture, women and land*, ed. Jean Davison, pp. 106–130. Boulder: Westview Press.

Schoonmaker-Freudenberger, Karen. 1991. L'integration en faveur des femmes et des enfants: Une évaluation des projets régionaux intègres soutenus par le Gouvernment du Sénégal et UNICEF. Dakar, Senegal.

Schroeder, Richard. 1991a. Of boycotts and bolongs: Vegetable marketing on the North Bank. Paper prepared for the National Workshop on Horticultural Programming in Rural Gambia. Banjul.

———. 1991b. The origins of the garden boom and its impact on rural Gambia. Paper prepared for the National Workshop on Horticultural Programming in Rural Gambia. Banjul.

———. 1991c. Defending a female cash crop: Threats to The Gambia's market garden boom. Paper prepared for the National Workshop on Horticultural Programming in Rural Gambia. Banjul.

———. 1993a. Shady practice: Gender and the political ecology of resource stabilization in Gambian garden / orchards. Ph.D. diss. University of California–Berkeley.

———. 1993b. Shady practice: Gender and the political ecology of resource stabilization in Gambian garden / orchards. *Economic Geography* 69 (4): 349–365.

———. 1995. Contradictions along the commodity road to environmental stabilization: Foresting Gambian gardens. *Antipode* 27 (4):325–342.

———. 1996. "Gone to their second husbands": Marital metaphors and conjugal contracts in The Gambia's female garden sector. *Canadian Journal of African Studies* 30 (1): 69–87.

———. 1999. Community, forestry and conditionality in The Gambia. *Africa* 69(1): 1–22.

———. Forthcoming. Geographies of environmental intervention in Africa. *Progress in Human Geography.*

Schroeder, Richard, and Roderick Neumann. 1995. Manifest ecological destinies: Local rights and global environmental agendas. *Antipode* 27 (4):321–324.

Schroeder, Richard, and Krisnawati Suryanata. 1996. Gender and class power in agroforestry: Case studies from Indonesia and West Africa. In *Liberation ecology: Environment, development, social movements,* eds. Richard Peet and Michael Watts, pp. 188–204. London: Routledge.

Schroeder, Richard, and Michael Watts. 1991. Struggling over strategies, fighting over food: Adjusting to food commercialization among Mandinka peasants in The Gambia. In *Research in rural sociology and development: Vol. 5, household strategies,* eds. Harry Schwarzweller and Daniel Clay, pp. 45–72. Greenwich, CT: JAI Press.

Scott, James. 1976. *The moral economy of the peasant: Rebellion and subsistence in Southeast Asia.* New Haven: Yale University Press.

———. 1985. *Weapons of the weak: Everyday forms of peasant resistance.* New Haven: Yale University Press.

———. 1990. *Domination and the arts of resistance: Hidden transcripts.* New Haven: Yale University Press.

Sen, Gita, and Caren Grown. 1987. *Development, crises and alternative visions: Third World women's perspectives.* New York: Monthly Review Press.

Shaikh, Asif, Eric Arnould, K. Christophersen, R. Hagen, J. Tabor, and Peter Warshall. 1988. *Opportunities for sustained development, successful natural resources management in the Sahel.* Vols. 1 and 2. Washington, DC: USAID.

Shipton, Parker. 1995. How Gambians save: Cultural and economic strategy at an ethnic crossroads. In *Money matters: Instability, values and social payments in the modern history of West African communities,* ed. Jane Guyer, pp. 245–276. Portsmouth, NH: Heinemann.

Shipton, Parker, and Mitzi Goheen. 1992. Understanding African land-holding: Power wealth and meaning. *Africa* 62 (3):307–326.

Shiva, Vandana. 1988. *Staying alive: Women, ecology and development.* Atlantic Highlands, NJ: Zed Books.

Smith, Neil. 1996. The production of nature. In *FutureNatural: Nature / Science / Culture,* eds. George Robertson, Melinda Mash, Lisa Tickner, Jon Bird, Barry Curtis, and Tim Putnam, pp. 35–54. London: Routledge.

Stamp, Patricia. 1990. *Technology, gender and power in Africa.* Ottawa: IDRC.

Stichter, Sharon, and Jane Parpart, eds. 1988. *Patriarchy and class: African women in the home and the workforce.* Boulder: Westview Press.

Stone, M. Priscilla, Glenn Stone, and Robert Netting. 1995. The sexual division of labor in Kofyar agriculture. *American Ethnologist* 22 (1):165–186.

Suryanata, Krisnawati. 1994. Fruit trees under contract: Tenure and land use change in upland Java. *World Development* 22 (10):1567–1578.

Swindell, Kenneth. 1980. Serawoolies, tillibunkas and strange farmers: The development of migrant groundnut farming along the Gambia River, 1848–1895. *Journal of African History* 21 (1):93–104.

Szasz, Andrew. 1994. *Ecopopulism: Toxic waste and the movement for environmental justice.* Minneapolis: University of Minnesota Press.

TANGO (The Association of NGOs). N.d. The association of non-governmental organizations (TANGO) register. Banjul, The Gambia.

Taylor, Peter. and Fred Buttel. 1992. How do we know we have environmental problems? *Geoforum* 23:405–416.

Thiesen, Albert, S. Jallow, John Nittler, and Dominique Philippon. 1989. African food systems initiative, project document. U.S. Peace Corps, The Gambia.

Thoma, Wolfgang. 1989. Possibilities of introducing community forestry in The Gambia, part I. Gambia-German Forestry Project, Deutsche Gesellschaft for Technische Zusammenarbeit (GTZ). Feldkirchen, Germany: Deutsche Forstservice.

Thoma, Wolfgang. and S. Jaiteh. 1991. Possibilities of introducing community forestry in The Gambia, part III. Gambia-German Forestry Project, Deutsche Gesellschaft for Technische Zusammenarbeit (GTZ). Feldkirchen, Germany: Deutsche Forstservice.

Thomas-Slayter, Barbara, and Dianne Rocheleau. 1995. *Gender, environment, and development in Kenya: A grassroots perspective.* Boulder: Lynne Reinner.

UNESCO Courier. 1992. Women speak out on the environment (special issue). March.

United Nations. 1984a. Africa's food and agricultural situation reviewed in second committee. *United Nations Chronicle* 21 (10–11): 58–60.

———. 1984b. Assembly adopts declaration on Africa, calls for strategies to aid continent. *United Nations Chronicle* 21 (10–11):7–10.

———. 1984c. Most speakers voice alarm over malnutrition, famine ravaging Africa. *United Nations Chronicle* 21 (10–11):11–15.

———. 1985a. $1.7 billion in new aid needed for drought-stricken nations. *United Nations Chronicle* 22 (2):6–7.

———. 1985b. More than 1 million African drought victims saved through international efforts, OEOA says. *United Nations Chronicle* 22 (8):50.

———. 1986. General Assembly approves call for special Assembly session on critical economic situation in Africa. *United Nations Chronicle* 23 (2): 32–36.

———. 1987. Appeal made for urgent and intensified international efforts to meet Africa's emergency needs. *United Nations Chronicle* 24 (1):47.

USAID. 1991. Program assistance initial proposal: Agricultural and natural resource program. Banjul, The Gambia.

———. 1992. Program assistance approval document. Agriculture and natural resources (ANR) program and agriculture and natural resources (ANR) support project. Banjul, The Gambia.

———. 1993. *Africa: Growth renewed, hope rekindled, a report on the performance of the development fund for Africa 1988–1992.* Washington, DC: USAID.

Vakis, Nicos. 1986. Post harvest handling and quality control for export development of fresh horticultural produce. Regional (Africa) market news service for selected products and assistance aiming at achieving co-ordinated export development. Gambia. UNCTAD / GATT.

Vestal, Theodore. 1985. Famine in Ethiopia: Crisis of many dimensions. *Africa Today* 32 (4):7–28.

von Braun, Joachim, Ken Johm, Sambou Kinteh, and Detlev Puetz. 1990. *Structural adjustment, agriculture and nutrition: Policy options in The Gambia.* Working Papers on Commercialization of Agriculture and Nutrition No. 4. Washington, DC: International Food Policy Research Institute (IFPRI).

von Braun, Joachim, and Patrick Webb. 1987. The impact of a new agricultural technology on intrahousehold division of labor in a West African setting. Unpublished manuscript. Washington, DC: International Food Policy Research Institute (IFPRI).

Watts, Michael. 1983a. On the poverty of theory: Natural hazards research in context. In *Interpretations of calamity,* ed. Kenneth Hewitt, pp. 231–262. Boston: Allen and Unwin.

———. 1983b. *Silent violence: Food, famine and peasantry in northern Nigeria.* Berkeley: University of California Press.

———. 1989. The agrarian crisis in Africa: Debating the crisis. *Progress in Human Geography* 13 (1):1–42.

———. 1993. Idioms of land and labor. In *Land tenure in Africa,* eds. Thomas Bassett and Donald Crummey, pp. 159–196. Madison: University of Wisconsin Press.

———. 1994. Development II: The privatization of everything? *Progress in Human Geography* 18 (3): 371–384.

Webb, Patrick. 1984. Of rice and men: The story behind The Gambia's decision to dam its river. In *The social and environmental effects of large dams*, vol. 2, eds. Edward Goldsmith and Nicholas Hildyard, pp. 120–130. Wadebridge, UK: Wadebridge Ecological Centre.

Weil, Peter. 1973. Wet rice, women and adaptation in The Gambia. *Rural Africana*, 19 (Winter): 20–29.

———. 1986. Agricultural intensification and fertility in The Gambia. In *Culture and reproduction: An anthropological critique of demographic transition theory*, ed. W. Penn Handwerker, pp. 294–320. Boulder: Westview Press.

Whitehead, Ann. 1981. "I'm hungry mum": The politics of domestic budgeting. In *Of marriage and the market*, eds. Kate Young, Carol Wolkowitz, and Roslyn McCullagh, pp. 88–111. London: CSE Books.

Wolf, Diane, ed. 1996. *Feminist dilemmas in fieldwork*. Boulder: Westview Press.

World Bank. 1995. *Mainstreaming the environment: The World Bank Group and the environment since the Rio Earth Summit*. Washington, DC: World Bank.

———. 1996. *Toward environmentally sustainable development in sub-Saharan Africa*. Washington, DC: World Bank.

Worldview International Foundation. 1990. *WIF Newsletter* 3 (1).

Yarbo, K., and Teresita Planas. 1990. Production and marketing linkage experiences—Western Divisions' women communal vegetable growing schemes. Paper presented to the National Horticultural Planning Workshop.

Zerner, Charles. Forthcoming. *People, plants and justice: Resource extraction and conservation in tropical developing countries*. New York: Columbia University Press.

Index